D0287655

QUEEN ELIZABETH

From the portrait in the National Portrait Gallery

Elizabeth *and* Essex

A Tragic History

by Lytton Strachey

HARCOURT, BRACE AND COMPANY
New York

TENTH PRINTING, JANUARY, 1930

PRINTED IN THE UNITED STATES OF AMERICA
BY THE QUINN & BODEN COMPANY, INC., RAHWAY, N. J.

TO
ALIX *and* JAMES
STRACHEY

ILLUSTRATIONS

ELIZABETH *and* ESSEX

I

THE English Reformation was not merely a religious event; it was also a social one. While the spiritual mould of the Middle Ages was shattered, a corresponding revolution, no less complete and no less far-reaching, occurred in the structure of secular life and the seat of power. The knights and ecclesiastics who had ruled for ages vanished away, and their place was taken by a new class of persons, neither chivalrous nor holy, into whose competent and vigorous hands the reins, and the sweets, of government were gathered. This remarkable aristocracy, which had been created by the cunning of Henry VIII, overwhelmed at last the power that had given it being. The figure on the throne became a shadow, while the Russells, the Cavendishes, the Cecils, ruled over England in supreme solidity. For many generations they *were* England; and it is difficult to imagine an England without them, even today.

The change came quickly—it was completed during the reign of Elizabeth. The rebellion of the Northern Earls in 1569 was the last great effort of the old dispensation to escape its doom. It failed; the wretched Duke of Norfolk—the feeble Howard who had dreamt of marrying Mary Queen of Scots—was beheaded; and the new social system was finally secure. Yet the spirit of the ancient feudalism was not quite exhausted. Once more, before the reign was over, it flamed up, embodied in a single individual—Robert Devereux, Earl of Essex. The flame was glorious—radiant with the colours of antique knighthood and the flashing gallantries of the past; but no substance fed it; flaring wildly, it tossed to and fro in the wind; it was suddenly put out. In the history of Essex, so perplexed in its issues, so desperate in its perturbations, so dreadful in its conclusion, the spectral agony of an abolished world is discernible through the tragic lineaments of a personal disaster.

His father, who had been created Earl of Essex by Elizabeth, was descended from all the great houses of medieval England. The Earl of Huntingdon, the Marquis of Dorset, the Lord Ferrers—Bohuns, Bourchiers, Rivers, Plantagenets—they crowded into his pedigree. One of his ancestresses, Eleanor de Bohun, was the sister of Mary, wife of Henry IV; another, Anne Woodville, was the sister of Elizabeth, wife of Edward IV; through Thomas of Woodstock, Duke of Gloucester, the family traced its descent from Edward III. The first Earl had been a man of dreams—virtuous and unfortunate. In the spirit of a crusader he had set out to subdue Ireland; but the intrigues of the Court, the economy of the Queen, and the savagery of the kerns had been too much for him, he had effected nothing, and had died at last a ruined and broken-hearted man. His son Robert was born in 1567. Nine years old when his father died, the boy found himself the inheritor of an illustrious name and the poorest Earl in England. But that was not all. The complex influences which shaped his destiny were present at his birth: his mother was as much a representative of the new nobility as his father of the old. Lettice Knollys's grandmother was a sister of Anne Boleyn; and thus Queen Elizabeth was Essex's first cousin twice removed. A yet more momentous relationship came into being when, two years after the death of the first Earl, Lettice became the wife of Robert Dudley, Earl of Leicester. The fury of her Majesty and the mutterings of scandal were passing clouds of small significance; what remained was the fact that Essex was the step-son of Leicester, the Queen's magnificent favourite, who, from the moment of her accession, had dominated her Court. What more could ambition ask for? All the ingredients were present—high birth, great traditions, Court influence, even poverty—for the making of a fine career.

The young Earl was brought up under the guardianship of Burghley. In his tenth year he was sent to Trinity College, Cambridge, where in 1581, at the age of 14, he received the degree of Master of Arts. His adolescence passed in the country, at one or other of his remote western estates—at Lanfey in Pembrokeshire, or, more often, at Chartley in Staffordshire, where the ancient house, with its carved timber, its embattled top, its windows enriched with the arms and devices of Devereux and Ferrers, stood romantically in the midst of the vast chase, through which the red deer and the fallow deer, the badger and the wild boar, ranged at will. The youth loved hunting and all the sports of manhood; but he loved reading too. He could write correctly in Latin and beautifully in English; he might have been a scholar, had he not been so spirited a nobleman. As he grew up this double nature seemed to be reflected in his physical complexion. The blood flew through his veins in vigorous vitality; he ran and tilted with the sprightliest; and then suddenly health would ebb away from him, and the pale boy would lie for hours in his chamber, obscurely melancholy, with a Virgil in his hand.

When he was eighteen, Leicester, sent with an army to the Netherlands, appointed him General of the Horse. The post was less responsible than picturesque, and Essex performed its functions perfectly. Behind the lines, in festive tournaments, "he gave all men great hope," says the Chronicler, "of his noble forwardness in arms"—a hope that was not belied when the real fighting came. In the mad charge of

Zutphen he was among the bravest, and was knighted by Leicester after the action.

More fortunate—or so it seemed—than Philip Sidney, Essex returned scatheless to England. He forthwith began an assiduous attendance at Court. The Queen, who had known him from his childhood, liked him well. His stepfather was growing old; in that palace a white head and a red face were serious handicaps; and it may well have seemed to the veteran courtier that the favour of a young connection would strengthen his own hand, and, in particular, counterbalance the rising influence of Walter Raleigh. Be that as it may, there was soon no occasion for pushing Essex forward. It was plain to all—the handsome, charming youth, with his open manner, his boyish spirits, his words and looks of adoration, and his tall figure, and his exquisite hands, and the auburn hair on his head, that bent so gently downwards, had fascinated Elizabeth. The new star, rising with extraordinary swiftness, was suddenly seen to be shining alone in the firmament. The Queen and the Earl were never apart. She was fifty-three, and he was not yet twenty: a dangerous concatenation of ages. Yet, for the moment—it was the May of 1587—all was smooth and well. There were long talks, long walks and rides through the parks and the woods round London, and in the evening there was more talk, and laughter, and then there was music, until, at last, the rooms at Whitehall were empty, and they were left, the two, playing cards together. On and on through the night they played—at cards or one game or another, so that, a contemporary gossip tells us, "my Lord cometh not to his own lodging till birds sing in the morning." Thus passed the May of 1587 and the June.

If only time could have stood still for a little and drawn out those halcyon weeks through vague ages of summer! The boy, in his excitement, walking home through the dawn, the smiling Queen in the darkness . . . but there is no respite for mortal creatures. Human relationships must either move or perish. When two consciousnesses come to a certain nearness the impetus of their interactions, growing ever intenser and intenser, leads on to an unescapable climax. The crescendo must rise to its topmost note; and only then is the pre-ordained solution of the theme made manifest.

II

THE reign of Elizabeth (1558 to 1603) falls into two parts:—the thirty years that preceded the defeat of the Spanish Armada, and the fifteen that followed it. The earlier period was one of preparation; it was then that the tremendous work was accomplished which made England a coherent nation, finally independent of the Continent, and produced a state of affairs in which the whole energies of the country could find free scope. During those long years the dominating qualities of the men in power were skill and prudence. The times were so hard that anything else was out of place. For a whole generation the vast caution of Burghley was the supreme influence in England. The lesser figures followed suit; and, for that very reason, a certain indistinctness veils them from our view. Walsingham worked underground; Leicester, with all his gorgeousness, is dim to us—an uncertain personage, bending to every wind; the Lord Chancellor Hatton danced, and that is all we know of him. Then suddenly the kaleidoscope shifted; the old ways, the old actors, were swept off with the wreckage of the Armada. Burghley alone remained—a monument from the past. In the place of Leicester and Walsingham, Essex and Raleigh—young, bold, coloured, brilliantly personal—sprang forward and filled the scene of public action. It was the same in every other field of national energy: the snows of the germinating winter had melted, and the wonderful spring of Elizabethan culture burst into life.

The age—it was that of Marlowe and Spenser, of the early Shakespeare and the Francis Bacon of the Essays—needs no description: everybody knows its outward appearances and the literary expressions of its heart. More valuable than descriptions, but what perhaps is unattainable, would be some means by which the modern mind might reach to an imaginative comprehension of those beings of three centuries ago—might move with ease among their familiar essential feelings—might touch, or dream that it touches (for such dreams are the stuff of history) the very "pulse of the machine." But the path seems closed to us. By what art are we to worm our way into those strange spirits, those even stranger bodies? The more clearly we perceive it, the more remote that singular universe becomes. With very few exceptions—possibly with the single exception of Shakespeare—the creatures in it meet us without intimacy; they are exterior visions, which we know, but do not truly understand.

It is, above all, the contradictions of the age that baffle our imagination and perplex our intelligence.

Human beings, no doubt, would cease to be human beings unless they were inconsistent; but the inconsistency of the Elizabethans exceeds the limits permitted to man. Their elements fly off from one another wildly; we seize them; we struggle hard to shake them together into a single compound, and the retort bursts. How is it possible to give a coherent account of their subtlety and their *naïveté*, their delicacy and their brutality, their piety and their lust? Wherever we look, it is the same. By what perverse magic were intellectual ingenuity and theological ingenuousness intertwined in John Donne? Who has ever explained Francis Bacon? How is it conceivable that the puritans were the brothers of the dramatists? What kind of mental fabric could that have been which had for its warp the habits of filth and savagery of sixteenth-century London and for its woof an impassioned familiarity with the splendour of Tamburlaine and the exquisiteness of Venus and Adonis? Who can reconstruct those iron-nerved beings who passed with rapture from some divine madrigal sung to a lute by a bewitching boy in a tavern to the spectacle of mauled dogs tearing a bear to pieces? Iron-nerved? Perhaps; yet the flaunting man of fashion, whose codpiece proclaimed an astonishing virility, was he not also, with his flowing hair and his jewelled ears, effeminate? And the curious society which loved such fantasies and delicacies—how readily would it turn and rend a random victim with hideous cruelty! A change of fortune—a spy's word—and those same ears might be sliced off, to the laughter of the crowd, in the pillory; or, if ambition or religion made a darker embroilment, a more ghastly mutilation—amid a welter of moral platitudes fit only for the nursery and dying confessions in marvellous English—might diversify a traitor's end.

It was the age of *baroque;* and perhaps it is the incongruity between their structure and their ornament that best accounts for the mystery of the Elizabethans. It is so hard to gauge, from the exuberance of their decoration, the subtle, secret lines of their inner nature. Certainly this was so in one crowning example—certainly no more *baroque* figure ever trod this earth than the supreme phenomenon of Elizabethanism—Elizabeth herself. From her visible aspect to the profundities of her being, every part of her was permeated by the bewildering discordances of the real and the apparent. Under the serried complexities of her raiment—the huge hoop, the stiff ruff, the swollen sleeves, the powdered pearls, the spreading, gilded gauzes—the form of the woman vanished, and men saw instead an image—magnificent, portentous, self-created—an image of regality, which yet, by a miracle, was actually alive. Posterity has suffered by a similar deceit of vision. The great Queen of its imagination, the lion-hearted heroine, who flung back the insolence of Spain and crushed the tyranny of Rome with splendid unhesitating gestures, no more resembles the Queen of fact than the clothed Elizabeth the naked one. But, after all, posterity is privileged. Let us draw nearer; we shall do no wrong now to that Majesty, if we look below the robes.

The lion heart, the splendid gestures—such heroic things were there, no doubt—visible to everybody; but their true significance in the general scheme of her character was remote and complicated. The sharp and hostile eyes of the Spanish ambassadors saw something different; in their opinion, the outstanding characteristic of Elizabeth was pusillanimity. They were wrong; but they perceived more of the truth than the idle onlooker. They had come into contact with those forces in the Queen's mind which proved, incidentally, fatal to themselves, and brought her, in the end, her enormous triumph. That triumph was not the result of heroism. The very contrary was the case: the grand policy which dominated Elizabeth's life was the most unheroic conceivable; and her true history remains a standing lesson for melodramatists in statecraft. In reality, she succeeded by virtue of all the qualities which every hero should be without—dissimulation, pliability, indecision, procrastination, parsimony. It might almost be said that the heroic element chiefly appeared in the unparalleled lengths to which she allowed those qualities to carry her. It needed a lion heart indeed to spend twelve years in convincing the world that she was in love with the Duke of Anjou, and to stint the victuals of the men who defeated the Armada; but in such directions she was in very truth capable of everything. She found herself a sane woman in a universe of violent maniacs, between contending forces of terrific intensity—the rival nationalisms of France and Spain, the rival religions of Rome and Calvin; for years it had seemed inevitable that she should be crushed by one or other of them, and she had survived because she had been able to meet the extremes around her with her own extremes of cunning and prevarication. It so happened that the subtlety of her intellect was exactly adapted to the complexities of her environment. The balance of power between France and Spain, the balance of factions in France and Scotland, the swaying fortunes of the Netherlands, gave scope for a tortuosity of diplomacy which has never been completely unravelled to this day. Burghley was her chosen helper, a careful steward after her own heart; and more than once Burghley gave up the puzzle of his mistress's proceedings in despair. Nor was it only her intellect that served her; it was her temperament as well. That too—in its mixture of the masculine and the feminine, of vigour and sinuosity, of pertinacity and vacillation—was precisely what her case required. A deep instinct made it almost impossible for her to come to a fixed determination upon any subject whatever. Or, if she did, she immediately proceeded to contradict her resolution with the utmost violence, and, after that, to contradict her contradiction more violently still. Such was her nature—to float, when it was calm, in a sea of indecisions, and, when the wind rose, to tack hectically from side to side. Had it been otherwise—had she possessed, according to the approved pattern of the strong man of action, the capacity for taking a line and sticking to it—she would have been lost. She would have become inextricably entangled in the forces that surrounded her, and, almost inevitably, swiftly destroyed. Her femininity saved her. Only a woman could have shuffled so shamelessly, only a woman could have abandoned with such unscrupulous completeness the last shreds not only of consistency, but of dignity, honour, and common decency, in order to escape the appalling necessity of having, really and truly, to make up her mind. Yet it is true that a woman's evasiveness was not enough; male courage, male energy were needed, if she were to escape the pressure that came upon her from every side. Those qualities she also possessed; but their value to her—it was the final paradox of her career—was merely that they made her strong enough to turn her back, with an indomitable persistence, upon the ways of strength.

Religious persons at the time were distressed by her

conduct, and imperialist historians have wrung their hands over her since. Why could she not suppress her hesitations and chicaneries and take a noble risk? Why did she not step forth, boldly and frankly, as the leader of Protestant Europe, accept the sovereignty of Holland, and fight the good fight to destroy Catholicism and transfer the Spanish Empire to the rule of England? The answer is that she cared for none of those things. She understood her true nature and her true mission better than her critics. It was only by an accident of birth that she was a Protestant leader; at heart she was profoundly secular; and it was her destiny to be the champion, not of the Reformation, but of something greater—the Renaissance. When she had finished her strange doings, there was civilisation in England. The secret of her conduct was, after all, a simple one: she had been gaining time. And time, for her purposes, was everything. A decision meant war—war, which was the very antithesis of all she had at heart. Like no other great statesman in history, she was, not only by disposition, but in practice, pacific. It was not that she was much disturbed by the cruelty of war—she was far from sentimental; she hated it for the best of all reasons—its wastefulness. Her thrift was spiritual as well as material, and the harvest that she gathered in was the great Age, to which, though its supreme glories were achieved under her successor, her name has been rightly given. For without her those particular fields could never have come to ripeness; they would have been trodden down by struggling hordes of nationalists and theologians. She kept the peace for thirty years—by dint, it is true, of one long succession of disgraceful collapses and unheard-of equivocations; but she kept it, and that was enough for Elizabeth.

To put the day of decision off—and off—and off—it seemed her only object, and her life passed in a passion of postponement. But here, too, appearances were deceitful, as her adversaries found to their cost. In the end, when the pendulum had swung to and fro for ages, and delay had grown grey, and expectation sunk down in its socket . . . something terrible happened. The crafty Maitland of Lethington, in whose eyes the God of his fathers was "ane bogle of the nursery," declared with scorn that the Queen of England was inconstant, irresolute, timorous, and that before the game was played out he would "make her sit upon her tail and whine, like ane whippet hound." Long years passed, and then suddenly the rocks of Edinburgh Castle ran down like sand at Elizabeth's bidding, and Maitland took refuge from the impossible ruin in a Roman's death. Mary Stuart despised her rival with a virulent French scorn; and, after eighteen years, at Fotheringay, she found she was mistaken. King Philip took thirty years to learn the same lesson. For so long had he spared his sister-in-law; but now he pronounced her doom; and he smiled to watch the misguided woman still negotiating for a universal peace, as his Armada sailed into the Channel.

Undoubtedly there was a touch of the sinister about her. One saw it in the movements of her extraordinarily long hands. But it was a touch and no more—just enough to remind one that there was Italian blood in her veins—the blood of the subtle and cruel Visconti. On the whole, she was English. On the whole, though she was infinitely subtle, she was not cruel; she was almost humane for her times; and her occasional bursts of savagery were the results of fear or temper. In spite of superficial resemblances, she was the very opposite of her most dangerous enemy—the weaving spider of the Escurial. Both were masters of dissimulation and lovers of delay; but the leaden foot of Philip was the symptom of a dying organism, while Elizabeth temporised for the contrary reason—because vitality can afford to wait. The fierce old hen sat still, brooding over the English nation, whose pullulating energies were coming swiftly to ripeness and unity under her wings. She sat still; but every feather bristled; she was tremendously alive. Her super-abundant vigour was at once alarming and delightful. While the Spanish ambassador declared that ten thousand devils possessed her, the ordinary Englishman saw in King Hal's full-blooded daughter a Queen after his own heart. She swore; she spat; she struck with her fist when she was angry; she roared with laughter when she was amused. And she was often amused. A radiant atmosphere of humour coloured and softened the harsh lines of her destiny, and buoyed her up along the zigzags of her dreadful path. Her response to every stimulus was immediate and rich: to the folly of the moment, to the clash and horror of great events, her soul leapt out with a vivacity, an abandonment, a complete awareness of the situation, which made her, which makes her still, a fascinating spectacle. She could play with life as with an equal, wrestling with it, making fun of it, admiring it, watching its drama, intimately relishing the strangeness of circumstance, the sudden freaks of fortune, the perpetual unexpectedness of things. "Per molto variare la natura è bella" was one of her favourite aphorisms.

The variations in her own behaviour were hardly less frequent than nature's. The rough hectoring dame with her practical jokes, her out-of-doors manners, her passion for hunting, would suddenly become a stern-faced woman of business, closeted for long hours with secretaries, reading and dictating despatches, and examining with sharp exactitude the minutiae of accounts. Then, as suddenly, the cultivated lady of the Renaissance would shine forth. For Elizabeth's accomplishments were many and dazzling.

She was mistress of six languages besides her own, a student of Greek, a superb calligraphist, an excellent musician. She was a connoisseur of painting and poetry. She danced, after the Florentine style, with a high magnificence that astonished beholders. Her conversation, full, not only of humour, but of elegance and wit, revealed an unerring social sense, a charming delicacy of personal perception. It was this spiritual versatility which made her one of the supreme diplomatists of history. Her protean mind, projecting itself with extreme rapidity into every sinuous shape conceivable, perplexed the most clear-sighted of her antagonists and deluded the most wary. But her crowning virtuosity was her command over the resources of words. When she wished, she could drive in her meaning up to the hilt with hammer blows of speech, and no one ever surpassed her in the elaborate confection of studied ambiguities. Her letters she composed in a regal mode of her own, full of apophthegm and insinuation. In private talk she could win a heart by some quick felicitous *brusquerie;* but her greatest moments came when, in public audience, she made known her wishes, her opinions, and her meditations to the world. Then the splendid sentences, following one another in a steady volubility, proclaimed the curious workings of her intellect with enthralling force; while the woman's inward passion vibrated magically through the loud high uncompromising utterance and the perfect rhythms of her speech.

Nor was it only in her mind that these complicated contrasts were apparent; they dominated her physical being too. The tall and bony frame was subject to strange weaknesses. Rheumatisms racked her; intolerable headaches laid her prone in agony; a hideous ulcer poisoned her existence for years. Though her serious illnesses were few, a long succession of minor maladies, a host of morbid symptoms, held her contemporaries in alarmed suspense and have led some modern searchers to suspect that she received from her father an hereditary taint. Our knowledge, both of the laws of medicine and of the actual details of her disorders, is too limited to allow a definite conclusion; but at least it seems certain that, in spite of her prolonged and varied sufferings, Elizabeth was fundamentally strong. She lived to be seventy—a great age in those days—discharging to the end the laborious duties of government; throughout her life she was capable of unusual bodily exertion; she hunted and danced indefatigably; and—a significant fact, which is hardly compatible with any pronounced weakness of physique—she took a particular pleasure in standing up, so that more than one unfortunate ambassador tottered from her presence, after an audience of hours, bitterly complaining of his exhaustion. Probably the solution of the riddle—suggested at the time by various onlookers, and accepted by learned authorities since—was that most of her ailments were of an hysterical origin. That iron structure was a prey to nerves. The hazards and anxieties in which she passed her life would have been enough in themselves to shake the health of the most vigorous; but it so happened that, in Elizabeth's case, there was a special cause for a neurotic condition: her sexual organisation was seriously warped.

From its very beginning her emotional life had been subjected to extraordinary strains. The intensely impressionable years of her early childhood had been for her a period of excitement, terror, and tragedy. It is possible that she could just remember the day when, to celebrate the death of Catherine of Aragon, her father, dressed from top to toe in yellow, save for one white plume in his bonnet, led her to mass in a triumph of trumpets, and then, taking her in his arms, showed her to one after another of his courtiers, in high delight. But it is also possible that her very earliest memory was of a different kind: when she was two years and eight months old, her father cut off her mother's head. Whether remembered or no, the reactions of such an event upon her infant spirit must have been profound. The years that followed were full of trouble and dubiety. Her fate varied incessantly with the complex changes of her father's politics and marriages; alternately caressed and neglected, she was the heir to England at one moment and a bastard outcast the next. And then, when the old King was dead, a new and dangerous agitation almost overwhelmed her. She was not yet fifteen, and was living in the house of her stepmother, Catherine Parr, who had married the Lord Admiral Seymour, brother of Somerset, the Protector. The Admiral was handsome, fascinating and reckless; he amused himself with the Princess. Bounding into her room in the early morning, he would fall upon her, while she was in her bed or just out of it, with peals of laughter, would seize her in his arms and tickle her, and slap her buttocks, and crack a ribald joke. These proceedings continued for several weeks, when Catherine Parr, getting wind of them, sent Elizabeth to live elsewhere. A few months later Catherine died, and the Admiral proposed marriage to Elizabeth. The ambitious charmer, aiming at the supreme power, hoped to strengthen himself against his brother by a union with the royal blood. His plots were discovered; he was flung into the Tower, and the Protector sought to inculpate Elizabeth in the conspiracy. The agonised girl kept her head. The looks and the ways of Thomas Seymour had delighted her; but she firmly denied that she had ever contemplated marriage without the Protector's consent. In a masterly letter, written in an exquisite hand, she rebutted Somerset's charges. It was rumoured, she told him, that she was "with child by my Lord Admiral"; this was a "shameful schandler"; and she begged to be allowed to go to court, where all would see that it was so. The Protector found that he could do nothing with his fifteen-year-old antagonist; but he ordered the Admiral to be beheaded.

Such were the circumstances—both horrible and singular—in which her childhood and her puberty were passed. Who can wonder that her maturity should have been marked by signs of nervous infirmity? No sooner was she on the throne than a strange temperamental anomaly declared itself. Since the Catholic Mary Stuart was the next heir, the Protestant cause in England hung suspended, so long as Elizabeth remained unmarried, by the feeble thread of her life. The obvious, the natural, the inevitable conclusion was that the Queen's marriage must immediately take place. But the Queen was of a different opinion. Marriage was distasteful to her, and marry she would not. For more than twenty years, until age freed her from the controversy, she resisted, through an incredible series of delays, ambiguities, perfidies, and tergiversations, the incessant pressure of her ministers, her parliaments, and her people. Considerations of her own personal safety were of no weight with her. Her childlessness put a premium upon her murder; she knew it, and she smiled. The world was confounded by such unparalleled conduct. It was not as if an icy chastity possessed the heart of Elizabeth.

Far from it; the very opposite seemed to be the case. Nature had implanted in her an amorousness so irrepressible as to be always obvious and sometimes scandalous. She was filled with delicious agitation by the glorious figures of men. Her passion for Leicester dominated her existence from the moment when her sister's tyranny had brought them together in the Tower of London till the last hour of his life; and Leicester had virile beauty, and only virile beauty, to recommend him. Nor was Leicester alone in her firmament: there were other stars which, at moments, almost outshone him. There was the stately Hatton, so comely in a galliard; there was handsome Heneage; there was De Vere, the dashing king of the tiltyard; there was young Blount, with "his brown hair, a sweet face, a most neat composure, and tall in his person," and the colour that, when the eye of Majesty was fixed upon him, came and went so beautifully in his cheeks.

She loved them all; so it might be said by friends and enemies; for love is a word of questionable import; and over the doings of Elizabeth there hovered indeed a vast interrogation. Her Catholic adversaries roundly declared that she was Leicester's mistress, and had had by him a child, who had been smuggled away into hiding—a story that is certainly untrue. But there were also entirely contrary rumours afloat. Ben Jonson told Drummond, at Hawthornden, after dinner, that "she had a membrana on her, which made her uncapable of man, though for her delight she tryed many." Ben's loose talk, of course, has no authority; it merely indicates the gossip of the time; what is more important is the considered opinion of one who had good means of discovering the truth—Feria, the Spanish ambassador. After making careful inquiries, Feria had come to the conclusion, he told King Philip, that Elizabeth would have no children: "entiendo que ella no terna hijos," were his words. If this was the case, or if Elizabeth believed it to be so, her refusal to marry becomes at once comprehensible. To have a husband and no child would be merely to lose her personal preponderance and gain no counterbalancing advantages; the Protestant succession would be no nearer safety, and she herself would be eternally vexed by a master. The crude story of a physical malformation may well have had its origin in a subtler, and yet no less vital, fact. In such matters the mind is as potent as the body. A deeply seated repugnance to the crucial act of intercourse may produce, when the possibility of it approaches, a condition of hysterical convulsion, accompanied, in certain cases, by intense pain. Everything points to the conclusion that such—the result of the profound psychological disturbances of her childhood—was the state of Elizabeth. "I hate the idea of marriage," she told Lord Sussex, "for reasons that I would not divulge to a twin soul." Yes; she hated it; but she would play with it nevertheless. Her intellectual detachment and her supreme instinct for the opportunities of political chicanery led her on to dangle the promise of her marriage before the eyes of the coveting world. Spain, France, and the Empire—for years she held them, lured by that impossible bait, in the meshes of her diplomacy. For years she made her mysterious organism the pivot upon which the fate of Europe turned. And it so happened that a contributing circumstance enabled her to give a remarkable verisimilitude to her game. Though, at the centre of her being, desire had turned to repulsion, it had not vanished altogether; on the contrary, the compensating forces of nature had redoubled its vigour elsewhere. Though the precious citadel itself was never to be violated, there were surrounding territories, there were outworks and bastions over which exciting battles might be fought, and which might even, at moments, be allowed to fall into the bold hands of an assailant. Inevitably, strange rumours flew. The princely suitors multiplied their assiduities; and the Virgin Queen alternately frowned and smiled over her secret.

The ambiguous years passed, and the time came at length when there could be no longer a purpose in marriage. But the Queen's curious temperament remained. With the approach of old age, her emotional excitements did not diminish. Perhaps, indeed, they actually increased; though here too there was a mystification. Elizabeth had been attractive as a girl; she remained for many years a handsome woman; but at last the traces of beauty were replaced by hard lines, borrowed colours, and a certain grotesque intensity. Yet, as her charms grew less, her insistence on their presence grew greater. She had been content with the devoted homage of her contemporaries; but from the young men who surrounded her in her old age she required—and received—the expressions of romantic passion. The affairs of State went on in a fandango of sighs, ecstasies, and protestations. Her prestige, which success had made enormous, was still further magnified by this transcendental atmosphere of personal worship. Men felt, when they came near her, that they were in a superhuman presence. No reverence was too great for such a divinity. A splendid young nobleman—so the story went—while bowing low before her, had given vent to an unfortunate sound, and thereupon, such was his horrified embarrassment, he had gone abroad and travelled for seven years before venturing to return to the presence of his Mistress. The policy of such a system was obvious; and yet it was by no means all policy. Her clear-sightedness, so tremendous in her dealings with outward circumstances, stopped short when she turned her eyes within. There her vision grew artificial and confused. It seemed as if, in obedience to a subtle instinct, she had succeeded in becoming one of the greatest of worldly realists by dint of concentrating the whole romance of her nature upon herself. The result was unusual. The wisest of rulers, obsessed by a preposterous vanity, existed in a universe that was composed entirely either of absurd, rose-tinted fantasies or the coldest and hardest of facts. There were no transitions—only opposites, juxtaposed. The extraordinary spirit was all steel one moment and all flutters the next. Once more her beauty had conquered, once more her fascinations had evoked the inevitable response. She eagerly absorbed the elaborate adorations of her lovers, and, in the same instant, by a final stroke of luck and cunning, converted them —like everything else she had anything to do with— into a paying concern.

That strange court was the abode of paradox and uncertainty. The goddess of it, moving in a nimbus of golden glory, was an old creature, fantastically dressed, still tall, though bent, with hair dyed red above her pale visage, long blackening teeth, a high domineering nose, and eyes that were at once deep-set and starting forward—fierce, terrifying eyes, in whose dark blue depths something frantic lurked— something almost maniacal. She passed on—the peculiar embodiment of a supreme energy; and Fate and Fortune went with her. When the inner door was closed, men knew that the brain behind the eyes was at work there, with the consummate dexterity of long-practised genius, upon the infinite complexities of European statecraft and the arduous government of a nation. From time to time a raucous sound was heard—a high voice, rating: an ambassador was being admonished, an expedition to the Indies forbidden, something determined about the constitution of the Church of England. The indefatigable figure emerged at last, to leap upon a horse, to gallop through the glades, and to return, well satisfied, for an hour with the virginals. After a frugal meal—the wing of a fowl, washed down with a little wine and water—Gloriana danced. While the viols sounded, the young men, grouped about her, awaited what their destiny might bring forth. Sometimes the Earl was absent, and then what might not be hoped for, from that quick susceptibility, that imperious caprice? The excited deity would jest roughly with one and another, and would end by summoning some strong-limbed youth to talk with her in an embrasure. Her heart melted with his flatteries, and, as she struck him lightly on the neck with her long fingers, her whole being was suffused with a lasciviousness that could hardly be defined. She was a woman—ah, yes! a fascinating woman!—but then, was she not also a virgin, and old? But immediately another flood of feeling swept upwards and engulfed her; she towered; she was something more—

she knew it; what was it? Was she a man? She gazed at the little beings around her, and smiled to think that, though she might be their Mistress in one sense, in another it could never be so—that the very reverse might almost be said to be the case. She had read of Hercules and Hylas, and she might have fancied herself, in some half-conscious day-dream, possessed of something of that pagan masculinity. Hylas was a page—he was before her . . . but her reflections were disturbed by a sudden hush. Looking round, she saw that Essex had come in. He went swiftly towards her; and the Queen had forgotten everything, as he knelt at her feet.

III

THE summer idyll passed smoothly on, until, in the hot days of July, there was a thunderstorm. While the Earl conversed with the Queen in her chamber, the Captain of the Guard stood outside the door on duty; and the Captain of the Guard was a gentleman with a bold face—Sir Walter Raleigh. The younger son of a West-country squire, the royal favour had raised him in a few years to wealth and power: patents and monopolies had been showered upon him; he had become the master of great estates in England and Ireland; he was warden of the stannaries, Lord-Lieutenant of Cornwall, a Knight, a Vice-Admiral; he was thirty-five—a dangerous and magnificent man. His splendid bearing, his enterprising spirit, which had brought him to this unexpected grandeur—whither would they lead him in the end? The Fates had woven for him a skein of mingled light and darkness; fortune and misfortune, in equal measure and in strange intensity, were to be his.

The first stroke of the ill-luck that haunted his life had been the appearance at Court of the youthful Essex. Just as Raleigh must have thought that the Queen's fancy was becoming fixed upon him, just as the decay of Leicester seemed to open the way to a triumphant future—at that very moment the old favourite's stepson had come upon the scene with his boyish fascinations and swept Elizabeth off her feet. Raleigh suddenly found himself in the position of a once all-conquering beauty whose charms are on the wane. The Queen might fling him three or four estates of beheaded conspirators, might give him leave to plant a colony in America, might even snuff at his tobacco and bite a potato with a wry face—all that was nothing: her heart, her person, were with Essex, on the other side of the door. He knotted his black eyebrows, and determined not to sink without a struggle. During a country visit at Lord Warwick's, he succeeded in disturbing Elizabeth's mind. Lady Warwick was a friend of Essex's sister, Lady Dorothy Perrott, who, owing to a clandestine marriage, had been forbidden to appear at Court, and the rash hostess, believing that the Queen's anger had abated, had invited Lady Dorothy, as well as her brother, to the house. Raleigh told Elizabeth that Lady Dorothy's presence was a sign of deliberate disrespect on the part of Essex; whereupon Elizabeth ordered Lady Dorothy to keep to her room. Essex understood what had happened and did not hesitate. After supper, alone with the Queen and Lady Warwick, he made a vehement expostulation, defended his sister, and declared (as he told a friend, in a letter written immediately afterwards) that Elizabeth had acted as she did "only to please that knave Raleigh, for whose sake I saw she would both grieve me and my love, and disgrace me in the eye of the world." Elizabeth, no less vehemently, replied. "It seemed she could not well endure anything to be spoken against Raleigh and taking hold of one word, *disdain*, she said there was no such cause why I should disdain him." This speech "did trouble me so much, that, as near as I could, I did describe unto her what he had been and what he was." The daring youth went further. "What comfort can I have," he exclaimed, "to give myself over to the service of a mistress that is in awe of such a man?" All this time, the Captain of the Guard was at his post. "I spake, what of grief and choler, as much against him as I could, and I think he, standing at the door, might very well hear the worst that I spoke of himself." But his high words were useless; the dispute grew sharper; and when the Queen, from defending Raleigh, went on to attack Essex's mother, Lady Leicester, whom she particularly disliked, the young man would hear no more. He would send his sister away, he said, though it was almost midnight and "for myself," he told the agitated Elizabeth, "I had no joy to be in any place, but loth to be near about her, when I knew my affection so much thrown down, and such a wretch as Raleigh so highly esteemed of her." To this the Queen made no answer, "but turned her away to my Lady Warwick," and Essex, flinging from the room, first despatched his sister from the house under an escort of armed retainers and then rode off himself to Margate, determined to cross the Channel and take a part in the Dutch war. "If I return," he wrote, "I will be welcomed home; if not, *una bella morire* is better than a disquiet life." But the Queen was too quick for him. Robert Carey, sent galloping after him, found him before he had taken ship and brought him back to Her Majesty. There was a reconciliation; the royal favour blazed forth again; and within a month or two Essex was Master of the Horse and Knight of the Garter.

Yet, though the cloud had vanished, the sky was subtly changed. A first quarrel is always an ominous thing. In the curious scene at Lord Warwick's, under the cover of jealousy and wounded affection, a suppressed distrust, almost a latent hostility had, for a moment, come to the surface. And there was more; Essex had discovered that, young as he was, he could upbraid the great Queen with impunity. Elizabeth had been angry, disagreeable, and unyielding in her defence of Raleigh, but she had not ordered those audacious protestations to stop; it had almost seemed that she liked them.

IV

THE Armada was defeated; Leicester was dead. A new world was opening for the young and the adventurous. It was determined, under Drake's auspices, to make a counter-attack on Spain, and an armament was prepared to raid Corunna, take possession of Lisbon, detach Portugal from Philip, and place Don Antonio, who laid claim to the kingdom, on the throne. Excitement, booty, glory, fluttered before the imagination of every soldier and of Essex among the rest; but the Queen forbade him to go. He was bold enough to ignore her orders, and, leaving London on horseback one Thursday evening, he arrived in Plymouth on Saturday morning—a distance of 220 miles. This time he was too quick for his mistress. Taking ship immediately, with a detachment of troops under the veteran Sir Roger Williams, he sailed for the coast of Spain. Elizabeth was furious; she despatched messenger after messenger to Plymouth, ordered pinnaces to search the Channel, and, in an enraged letter to Drake, fulminated against the unfortunate Sir Roger. "His offence," she wrote, "is in so high a degree that the same deserveth to be punished by death, which if you have not already done, then we will and command you that you sequester him from all charge and service and cause him to be safely kept, until you shall know our further pleasure therein, as you will answer for the contrary at your perils; for as we have authority to rule so we look to be obeyed." If Essex, she continued, "be now come into the company of the fleet, we straightly charge you that you do forthwith cause him to be sent hither in safe manner. Which, if you do not, you shall look to answer for the same to your smart; for these be no childish actions. Therefore consider well of your doings herein." But her threats and her commands were alike useless. Essex joined the main body of the expedition unhindered and took a brave part in the skirmishes and marches in which it ingloriously ended. It turned out to be easier to repel an invasion than to make one. Some Spanish ships were burnt, but the Portuguese did not rise, and Lisbon shut herself up against Don Antonio and the English. Into one of the gates of the town Essex, as a parting gesture, thrust his pike, "demanding aloud if any Spaniard mewed therein durst adventure forth in favour of his mistress to break a lance." There was no reply; and the expedition returned to England.

The young man soon made his peace with the Queen; even Sir Roger Williams was forgiven. The happy days of the Court returned with hunting, feasting, and jousting. Raleigh, with a shrug, went off to Ireland, to look after his ten thousand acres, and Essex was free from even the shadow of a rivalry. Or was Charles Blount a rival? The handsome boy had displayed his powers in the tiltyard to such purpose that Elizabeth had sent him a golden queen from her set of chessmen, and he had bound the trophy to his arm with a crimson ribbon. Essex, when he saw it, asked what it was, and, on being told, "Now I perceive," he exclaimed, "that every fool must have a favour." A duel followed in Marylebone fields and Essex was wounded. "By God's death!" said Elizabeth, when she heard of it, "it was fit that someone or other should take him down, and teach him better manners." She was delighted to think that blood had been shed over her beauty; but afterwards she insisted on the two young men making up their quarrel. She was obeyed, and Blount became one of the Earl's most devoted followers.

The stream of royal kindness flowed on, though occasionally there were odd shallows in it. Essex was extravagant; he was more than £20,000 in debt; and the Queen graciously advanced him £3000 to ease his necessities. Then suddenly she demanded immediate repayment. Essex begged for delay, but the reply was sharp and peremptory; the money—or its equivalent in land—must be handed over at once. In a pathetic letter, Essex declared his submission and devotion. "Now that your Majesty repents yourself," he wrote, "of the favour you thought to do me, I would I could, with the loss of all the land I have, as well repair the breach which your unkind answer hath made in my heart, as I can with the sale of one poor manor answer the sum which your Majesty takes of me. Money and land are base things, but love and kindness are excellent things, and cannot be measured but by themselves." Her Majesty admired the phrasing, but disagreed with the economics; and shortly afterwards the manor at Keyston in Huntingdonshire, "of mine ancient inheritance," as Essex told Burghley, "free from incumbrance, a great circuit of ground, in a very good soil," passed into the royal possession.

She preferred to be generous in a more remunerative way. She sold to Essex, for a term of years, the right to farm the customs on the sweet wines imported into the country—and he might make what he could out of it. He made a great deal—at the expense of the public; but he was informed that, when the lease expired, it might or might not be renewed as her Majesty thought fit.

He was lavish in the protestations of his worship—his adoration—his love. That convenient monosyllable, so intense and so ambiguous, was for ever on his lips and found its way into every letter—those elegant, impassioned, noble letters, which still exist, with their stiff, quick characters and those silken ties that were once loosened by the long fingers of Elizabeth. She read and she listened with a satisfaction so extraordinary, so unprecedented, that when one day she learned that he was married she was only enraged for a fortnight. Essex had made an impeccable choice —the widow of Sir Philip Sidney and the daughter of Sir Francis Walsingham; he was twenty-three, handsome, vigorous, with an earldom to hand on to posterity; even Elizabeth could not seriously object. She stormed and ramped; then remembered that the relations between herself and her servant were unique and had nothing to do with a futile domesticity. The fascinating bridegroom pursued and cajoled her with ardours as romantic as ever; and she felt that a queen could ignore a wife.

Soon enough an occasion arose for showing the world that to be the favourite of Elizabeth involved public duties as well as private delights. Henry IV of France, almost overpowered by the Catholic League and the Spaniards, appealed urgently to England for help. Elizabeth wavered for several months, and then reluctantly decided that Henry must be supported— but only with the absolute minimum of expenditure. She agreed that four thousand men should be sent to Normandy to act with the Huguenots; and Essex, who had done all he could to bring her to this resolution, now begged to be put in command of the force. Three times the Queen refused his entreaties; at last he knelt before her for two hours; still she refused—

then suddenly consented. The Earl went off in high feather, but discovered before very long that the command even of the smallest army needs something more than knight errantry. During the autumn and winter of 1591, difficulties and perplexities crowded upon him. He was hasty, rash and thoughtless. Leaving the main body of his troops, he galloped with a small escort through a hostile country to consult with the French King about the siege of Rouen and on his return was almost cut off by the Leaguers. The Council wrote from England upbraiding him with needlessly risking his life, with "trailing a pike like a common soldier," and with going a-hawking in districts swarming with the enemy. The Queen despatched several angry letters; everything annoyed her; she suspected Essex of incompetence and the French King of treachery; she was on the point of ordering the whole contingent home. Once more, as in the Portuguese expedition, it turned out that foreign war was a dreary and unprofitable business. Essex lost his favourite brother in a skirmish; he was agonised by the Queen's severity; his army dwindled, from death and desertion, to one thousand men. The English fought with reckless courage at Rouen; but the Prince of Parma, advancing from the Netherlands, forced Henry to raise the siege. The unfortunate young man, racked with ague, was overcome by a sudden despair. "Unkindness and sorrow," he told the Queen, "have broken both my heart and my wits." "I wish," he declared to one of his friends, "to be out of my prison, which I account my life." Yet his noble spirit soon re-asserted itself. His reputation was retrieved by his personal bravery. He challenged the Governor of Rouen to single combat—it was his one and only piece of strategy—amid general applause. The Queen, however, remained slightly cynical. The Governor of Rouen, she said, was merely a rebel, and she saw no occasion for the giving or receiving of challenges. But Essex, whatever the upshot of the expedition, would be romantic to the last; and, when the time came for him to return to England, he did so with a gesture of ancient chivalry. Standing on the shore of France before his embarkation, he solemnly drew his sword from its scabbard, and kissed the blade.

V

THE spring of youth was almost over; in those days, at the age of twenty-five, most men had reached a full maturity. Essex kept something of his boyishness to the end, but he could not escape the rigours of time, and now a new scene—a scene of peril and gravity appropriate to manhood—was opening before him.

The circumstances of a single family—it has happened more than once in English history—dominated the situation. William Cecil, Lord Burghley, who had filled, since the beginning of the reign, the position of Prime Minister, was over seventy; he could not last much longer; who would succeed him? He himself hoped that his younger son, Robert, might step into his place. He had brought him up with that end in view. The sickly, dwarfed boy had been carefully taught by tutors, had been sent travelling on the Continent, had been put into the House of Commons, had been initiated in diplomacy, and gently, persistently, at every favourable moment, had been brought before the notice of the Queen. Elizabeth's sharp eye, uninfluenced by birth or position, perceived that the little hunchback possessed a great ability. When

Walsingham died, in 1590, she handed over to Sir Robert Cecil the duties of his office; and the young man of twenty-seven became in fact, though not in name, her principal secretary. The title and emoluments might follow later—she could not quite make up her mind. Burghley was satisfied; his efforts had succeeded; his son's foot was planted firmly in the path of power.

But Lady Burghley had a sister, who had two sons—Anthony and Francis Bacon. A few years older than their cousin Robert, they were, like him, delicate, talented, and ambitious. They had started life with high hopes: their father had been Lord Keeper—the head of the legal profession; and their uncle was, under the Queen, the most important person in England. But their father died, leaving them no more than the small inheritance of younger sons; and their uncle, all-powerful as he was, seemed to ignore the claims of their deserts and their relationship. Lord Burghley, it appeared, would do nothing for his nephews. Why was this? To Anthony and Francis the explanation was plain: they were being sacrificed to the career of Robert; the old man was jealous of them—afraid of them; their capacities were suppressed in order that Robert should have no competitors. Nobody can tell how far this was the case. Burghley, no doubt, was selfish and wily; but perhaps his influence was not always as great as it seemed; and perhaps, also, he genuinely mistrusted the singular characters of his nephews. However that may be, a profound estrangement followed. The outward forms of respect and affection were maintained; but the bitter disappointment of the Bacons was converted into a bitter animosity, while the Cecils grew more suspicious and hostile every day. At last the Bacons decided to abandon their allegiance to an uncle who was worse than useless, and to throw in their lot with some other leader, who would appreciate them as they deserved. They looked round, and Essex was their obvious choice. The Earl was young, active, impressionable; his splendid personal position seemed to be there, ready to hand, waiting to be transformed into something more glorious still—a supreme political predominance. They had the will and the wit to do it. Their uncle was dropping into dotage, their cousin's cautious brain was no match for their combined intelligence. They would show the father and the son, who had thought to shuffle them into obscurity, that it is possible to be too grasping in this world and that it is sometimes very far from wise to quarrel with one's poor relations.

So Anthony at any rate thought—a gouty young invalid, splenetic and uncompromising; but the imaginations of Francis were more complicated. In that astonishing mind there were concealed depths and deceptive shallows, curiously intermingled and puzzling in the extreme to the inquisitive observer. Francis Bacon has been described more than once with the crude vigour of antithesis; but in truth such methods are singularly inappropriate to his most unusual case. It was not by the juxtaposition of a few opposites, but by the infiltration of a multitude of highly varied elements, that his mental composition was made up. He was no striped frieze; he was shot silk. The detachment of speculation, the intensity of personal pride, the uneasiness of nervous sensibility, the urgency of ambition, the opulence of superb taste—these qualities, blending, twisting, flashing together, gave to his secret spirit the subtle and glittering superficies of a serpent. A serpent, indeed, might well have been his chosen emblem—the wise, sinuous, dangerous creature, offspring of mystery and the beautiful earth. The music sounds, and the great snake rises, and spreads its hood, and leans and hearkens, swaying in ecstasy; and even so the sage Lord Chancellor, in the midst of some great sentence, some high intellectual confection, seems to hold his breath in a rich beatitude, fascinated by the deliciousness of sheer style. A true child of the Renaissance, his multiplicity was not merely that of mental accomplishment, but of life itself. His mind might move with joy among altitudes and theories, but the variegated savour of temporal existence was no less dear to him—the splendours of high living—the intricacies of court intrigue—the exquisiteness of pages—the lights reflected from small pieces of coloured glass. Like all the greatest spirits of the age, he was instinctively and profoundly an artist. It was this aesthetic quality which on the one hand inspired the grandeur of his philosophical conceptions and on the other made him one of the supreme masters of the written word. Yet his artistry was of a very special kind; he was neither a man of science nor a poet. The beauty of mathematics was closed to him, and all the vital scientific discoveries of the time escaped his notice. In literature, in spite of the colour and richness of his style, his genius was essentially a prose one. Intellect, not feeling, was the material out of which his gorgeous and pregnant sentences were made. Intellect! It was the common factor in all the variations of his spirit; it was the backbone of the wonderful snake.

Life in this world is full of pitfalls: it is dangerous to be foolish, and it is also dangerous to be intelligent; dangerous to others, and, no less, to oneself. "Il est bon, plus souvent qu'on ne pense," said the wise and virtuous Malesherbes, "de savoir ne pas avoir de l'esprit." But that was one of the branches of knowledge that the author of the *Advancement of Learning* ignored. It was impossible for Francis Bacon to imagine that any good could ever come of being simple-minded. His intellect swayed him too completely. He was fascinated by it, he could not resist it, he must follow wherever it led. Through thought, through action, on he went—an incredibly clever man. Through action even? Yes, for though the medley of human circumstance is violent and confused, assuredly one can find one's way through it to some purpose if only one uses one's wits. So thought the cunning artist; and smiling he sought to shape, with his subtle razor-blade, the crude vague blocks of passion and fact. But razors may be fatal in such contingencies; one's hand may slip; one may cut one's own throat.

The miserable end—it needs must colour our vision of the character and the life. But the end was implicit in the beginning—a necessary consequence of qualities that were innate. The same cause which made Bacon write perfect prose brought about his worldly and his spiritual ruin. It is probably always disastrous not to be a poet. His imagination, with all its magnificence, was insufficient: it could not see into the heart of things. And among the rest his own heart was hidden from him. His psychological acuteness, fatally external, never revealed to him the nature of his own desires. He never dreamt how intensely human he was. And so his tragedy was bitterly ironical, and a deep pathos invests his story. One wishes to turn away one's gaze from the unconscious traitor, the lofty-minded sycophant, the exquisite intelligence entrapped and strangled in the web of its own weaving.

"Although our persons live in the view of heaven, yet our spirits are included in the caves of our own complexions and customs, which minister unto us infinite errors and vain opinions." So he wrote; and so, perhaps, at last, he actually realised—an old man, disgraced, shattered, alone, on Highgate hill, stuffing a dead fowl with snow.

But all this was still far distant in the busy years of the early nineties, so rich with excitements and possibilities. The issues were simplified by the disgrace and imprisonment of Raleigh, whose amorous intrigue with Elizabeth Throgmorton, one of the maids of honour, had infuriated the Queen. The field was cleared for the two opposing factions: the new party of Essex and his followers—aggressive and adventurous—and the old party of the Cecils, entrenched in the strongholds of ancient power. This was the essence of the political situation till the close of the century; but it was complicated and confused both by compromises and by bitternesses, which were peculiar to the time. The party system was still undreamt-of; and the hostile forces which would be grouped today as Government and Opposition, then found themselves side by side in a common struggle to control the executive. When, early in 1593, Essex was sworn of the Privy Council, he became the colleague of his rivals. It was for the Queen to choose her counsellors. She would listen to one and then to another; she would shift, according to her adviser, from one policy to its direct contrary; it was a system of government after her own heart. Thus it was that she could enjoy to the full the delicious sense of ruling—could decide, with the plenitude of power, between momentous eventualities—and, by that very means, could contrive to keep up an endless balance and a marvellous marking of time. Her servants, struggling with each other for influence, remained her servants still. Their profound hostility could not divert them from their duty of working together for the Queen. There was no such thing as going temporarily out of office; one was either in office or one was nothing at all. To fail might mean death; but, until that came, the dangerous enemy whose success was one's annihilation, met one every day in the close companionship of the council table and the narrow inner circle of the court.

Very swiftly Essex, with the Bacons at his back, grew to be something more than a favourite and emerged as a minister and a statesman. The young man was taking himself seriously at last. He was never absent from the Council; and when the House of Lords was in session, he was to be seen in his place as soon as the business of the day began—at seven o'clock in the morning. But his principal activities were carried on elsewhere—in the panelled gallery and the tapestried inner chambers of Essex House—the great Gothic family residence which overlooked the

ROBERT DEVEREUX, EARL OF ESSEX

From the portrait at Woburn Abbey, by kind permission of the Duke of Bedford, K.G., K.B.E.

river from the Strand. There it was that Anthony Bacon, his foot swathed in hot flannels, plied his indefatigable pen. There it was that a great design was planned and carried into execution. The Cecils were to be beaten on their own chosen ground. The control of foreign affairs—where Burghley had ruled supreme for more than a generation—was to be taken from them; their information was to be proved inaccurate, and the policy that was based on it confuted and reversed. Anthony had no doubt that this could be done. He had travelled for years on the continent; he had friends everywhere; he had studied the conditions of foreign states, the intricacies of foreign diplomacy, with all the energy of his acute and restless mind. If his knowledge and intelligence were supported by the position and the wealth of Essex, the combination would prove irresistible. And Essex did not hesitate; he threw himself into the scheme with all his enthusiasm. A vast correspondence began. Emissaries were sent out, at the Earl's expense, all over Europe, and letters poured in, from Scotland, France, Holland, Italy, Spain, Bohemia, with elaborate daily reports of the sayings of princes, the movements of armies, and the whole complex development of international intrigue. Anthony Bacon sat at the centre, receiving, digesting, and exchanging news. The work grew and grew, and before long, such was the multiplicity of business, he had four young secretaries to help him, among whom were the ingenious Henry Wotton and the cynical Henry Cuffe. The Queen soon perceived that Essex knew what he was talking about, when there was a discussion on foreign affairs. She read his memoranda, she listened to his recommendations; and the Cecils found, more than once, that their carefully collected intelligence was ignored. Eventually a strange situation arose, characteristic of that double-faced age. Essex almost attained the position of an alternative Foreign Secretary. Various ambassadors—Thomas Bodley was one—came under his influence, and, while corresponding officially with Burghley, sent at the same time parallel and more confidential communications to Anthony Bacon. If the gain to the public service was doubtful, the gain to Essex was clear; and the Cecils, when they got wind of what was happening, began to realise that they must reckon seriously with the house in the Strand.

Francis Bacon's connection with Essex was not quite so close as his brother's. A barrister and a Member of Parliament, he had a career of his own; and he occupied his leisure with literary exercises and philosophical speculations. Yet he was in intimate contact with Essex House. The Earl was his patron, whom he held himself ready to assist in every way, whenever his help was needed—with advice, or the drafting of state papers, or the composition of some elaborate symbolic compliment, some long-drawn-out Elizabethan charade, for the entertainment of the Queen. Essex, seven years his junior, had been, from the first moment of their meeting, fascinated by the intellectual splendour of the elder man. His enthusiastic nature leapt out to welcome that scintillating wisdom and that profound wit. He saw that he was in the presence of greatness. He vowed that this astonishing being, who was devoting himself so generously to his service, should have a noble reward. The Attorney-Generalship fell vacant, and Essex immediately declared that Francis Bacon must have the post. He was young and had not yet risen far in his profession—but what of that? He deserved something even greater; the Queen might appoint whom she would, and if Essex had any influence, the right man, for once, should be given preferment.

The Attorney-Generalship was indeed a prize worth having, and to receive it from the hand of Essex would bring a peculiar satisfaction to Lord Burghley's nephew—it would show that he might come to honour without the aid of his uncle. Francis smiled; he saw a great career opening before his imagination—judgeships—high offices of state—might he not ere long be given, like his father before him, the keeping of the Great Seal of England? A peerage!—Verulam, St. Albans, Gorhambury—what resounding title should he take? "My manor of Gorhambury"—the phrase rolled on his tongue; and then his chameleon mind took on another colour; he knew that he possessed extraordinary administrative capacity; he would guide the destinies of his country, the world should know his worth. But those, after all, were but small considerations. Most could be politicians, many could be statesmen; but might there not be reserved for him alone a more magnificent fate? To use his place and his power for the dissemination of learning, for the creation of a new and mighty knowledge, for a vast beneficence, spreading in ever wider and wider circles through all humanity . . . these were glorious ends indeed! As for himself—and yet another tint came over his fancy—that office would be decidedly convenient. He was badly in want of cash. He was extravagant; he knew it—it could not be helped. It was impossible for him to lead the narrow life of mean economies that poverty dictated. His exuberant temperament demanded the solace of material delights. Fine clothes were a necessity—and music—and a household with a certain state. His senses were fastidious; the smell of ordinary leather was torture to him, and he put all his servants into Spanish leather boots. He spent infinite trouble in obtaining a particular kind of small beer, which was alone tolerable to his palate. His eye—a delicate, lively hazel eye—"it was like the eye of a viper," said William Harvey—required the perpetual refreshment of beautiful things.

A group of handsome young men—mere names now—a Jones, a Percy—he kept about him, half servants and half companions, and he found in their equivocal society an unexpected satisfaction. But their high living added alarmingly to the expenses of his establishment. He was already in debt, and his creditors were growing disagreeable. There could be no doubt about it; to be made Attorney-General would be a supreme piece of good fortune, from every point of view.

Essex at first had little doubt that he would speedily obtain the appointment. He found the Queen in good humour; he put forward Bacon's name, and immediately discovered that a serious obstacle stood in the way of his desire. By an unlucky chance, a few weeks previously Bacon, from his place in the House of Commons, had opposed the granting of a subsidy which had been asked for by the Crown. The tax, he declared, was too heavy, and the time allowed for the levying of it too short. The House of Lords had intervened, and attempted to draw the Commons into a conference; whereupon Bacon had pointed out the danger of allowing the Lords to have any share in a financial discussion, with the result that their motion had been dropped. Elizabeth was very angry; interference in such a question from a member of the House of Commons appeared to her to be little short of disloyalty; and she forbade Bacon to appear before her. Essex tried to soften her in vain. Bacon's apologies, she considered, were insufficient—he had defended himself by asserting that he had done what he had merely from a sense of duty. He had, in fact, acted with a singular spirit; but it was for the last time. His speech against the subsidy had been extremely clever, but not to have made it would have been cleverer still. Never again would he be so ingenuous as to appear to be independent of the Court. The result of such plain dealing was all too obvious. The more Essex pressed his suit, the more objections the Queen raised. Bacon, she said, had had too little practice; he was a man of theory; and Edward Coke was a sounder lawyer. Weeks passed, months passed, and still the Attorney-Generalship hung in the wind, and the regeneration of mankind grew dubious amid a mountain of unpaid bills.

Essex continued sanguine; but Bacon perceived that if the delay lasted much longer he would be ruined. He raised money wherever he could. Anthony sold an estate, and gave him the proceeds. He himself determined to sell land; but only one property was available, and that he could not dispose of without the consent of his mother. Old Lady Bacon was a terrific dowager, who lived, crumpled and puritanical, in the country. She violently disapproved of her son Francis. She disapproved; but, terrific as she was, she found it advisable not to express her sentiments directly. There was something about her son Francis which made even her think twice before she displeased him. She preferred to address herself to Anthony on such occasions, to pour out her vexation before his less disquieting gaze, and to hope that some of it would reach the proper quarter. When she was approached by the brothers about the land, her fury rose to boiling-point. She wrote a long, crabbed, outraged letter to Anthony. She was asked, she said, to consent to the selling of property in order to pay for the luxurious living of Francis and his disreputable retainers. "Surely," she wrote, "I pity your brother, yet so long as he pitieth not himself but keepeth that bloody Percy, as I told him then, yea as a coach companion and bed companion—a proud, profane, costly fellow, whose being about him I verily fear the Lord God doth mislike and doth less bless your brother in credit and otherwise in his health surely I am utterly discouraged. . . . That Jones never loved your brother indeed, but for his own credit, living upon your brother, and thankless though bragging. . . . It is most certain that till first Enny, a filthy wasteful knave, and his Welshmen one after another—for take one and they will still swarm ill-favouredly—did so lead him in a train, he was a towardly young gentleman, and a son of much good hope in godliness." So she fulminated. She would only release the land, she declared, on condition that she received a complete account of Francis's debts and was allowed a free hand in the payment of them. "For I will not," she concluded, "have his cormorant seducers and instruments of Satan to him committing foul sin by his countenance, to the displeasing of God and his godly fear."

When this was handed on to Francis, he addressed to his mother an elaborate letter of protest and conciliation. She returned it to Anthony in a rage. "I send herein your brother's letter. Construe the interpretation. I do not understand his enigmatical folded writing." Her son, she said, had been blessed with "good gifts of natural wit and understanding. But the same good God that hath given them to him will I trust and heartily pray to sanctify his heart by the right use of them, to glorify the Giver of them to his own inward comfort." Her prayer—it is the common fate of the prayers of mothers—was only ironically answered. As for the land, old Lady Bacon found herself in the end no match for her two sons; she yielded without conditions; and Francis, for the time at least, was freed from his embarrassment.

Meanwhile Essex did not relax his efforts with the Queen. "I cannot tell," wrote Anthony to his mother, "in what terms to acknowledge the desert of the Earl's unspeakable kindness towards us both, but namely to him now at a pinch, which by God's help shortly will appear by good effects." In several long conferences, the gist of which, when they were over, he immediately reported by letter to one or other of the brothers, Essex urged Elizabeth to make the desired appointment. But the "good effects" were slow in coming. The vacancy had occurred in the April of 1593, and now the winter was closing in, and still it was unfilled. The Queen, it was clear, was giving yet another exhibition of her delaying tactics. During the repeated discussions with Essex about the qualifications of his friend, she was in her element. She raised every kind of doubt and difficulty, to every reply she at once produced a rejoinder, she suddenly wavered and seemed on the brink of a decision, she postponed everything on some slight pretext, she flew into a temper, she was charming, she danced off. Essex, who could not believe that he would fail, grew sometimes himself more seriously angry. The Queen was the more pleased. She pricked him with the pins of her raillery, and watched the tears of irritation starting to his eyes. The Attorney-Generalship and the fate of Francis Bacon had become entangled in the web of that mysterious amour. At moments flirtation gave way to passion. More than once that winter, the young man, suddenly sulky, disappeared, without a word of warning, from the Court. A blackness and a void descended upon Elizabeth; she could not conceal her agitation; and then, as suddenly, he would return, to be overwhelmed with scornful reproaches and resounding oaths.

The quarrels were short, and the reconciliations were delicious. On Twelfth Night there was acting and dancing at Whitehall. From a high throne, sumptuously decorated, the Queen watched the ceremonies, while beside her stood the Earl, with whom "she often devised in sweet and favourable manner." So the scene was described by Anthony Standen, an old courtier, in a letter that has come down to us. It was an hour of happiness and peace; and, amid the jewels and the gilded hangings, the incredible Princess, who had seen her sixtieth birthday, seemed to shine with an almost youthful glory. The lovely knight by her side had wrought the miracle—had smiled the long tale of hideous years into momentary nothingness. The courtiers gazed in admiration, with no sense of incongruity. "She was as beautiful," wrote Anthony Standen, "to my old sight, as ever I saw her."

Was it possible that to the hero of such an evening anything could be refused? If he had set his heart on the Attorney-Generalship for Bacon, surely he would have it. The time of decision seemed to be approaching. Burghley begged the Queen to hesitate no longer, and he advised her to give the place to Edward Coke. The Cecils believed that she would do so; and Sir Robert, driving with the Earl one day in a coach through the city, told him that the appointment would be made in less than a week. "I pray your

Lordship," he added, "to let me know whom you will favour." Essex replied that Sir Robert must surely be aware that he stood for Francis Bacon. "Lord!" replied Sir Robert, "I wonder your lordship should go about to spend your strength in so unlikely or impossible a manner. If your lordship had spoken of the *solicitorship*, that might be of easier digestion to her Majesty." At that Essex burst out. "Digest me no digestions," he cried; "for the attorneyship for Francis is that I must have. And in that I will spend all my power, might, authority, and amity, and with tooth and nail defend and procure the same for him against whomsoever; and whosoever getteth this office out of my hands for any other, before he have it, it shall cost him the coming by. And this be you assured of, Sir Robert; for now do I fully declare myself. And for your own part, Sir Robert, I think strange both of my lord Treasurer and you that you can have the mind to seek the preference of a stranger before so near a kinsman." Sir Robert made no reply; and the coach rattled on, with its burden of angry ministers. Henceforth there was no concealment; the two parties faced each other fiercely; they would try their strength over Coke and Bacon.

But Elizabeth grew more ambiguous than ever. The week passed, and there was no sign of an appointment. To make any decision upon any subject at all had become loathsome to her. She lingered in a spiritual palsy at Hampton Court; she thought she would go to Windsor; she gave orders to that effect, and countermanded them. Every day she changed her mind: it was impossible for her to determine even whether she wanted to move or to stay still. The whole Court was in an agony, half packed up. The carter in charge of the wagons in which the royal belongings were carried had been summoned for the third time, and for the third time was told that he might go away. "Now I see," he said, "that the Queen is a woman as well as my wife." The Queen, who was standing at a window, overheard the remark, and burst out laughing. "What a villain is this!" she said, and sent him three angels to stop his mouth. At last she did move—to Nonesuch. A few more weeks passed. It was Easter, 1594. She suddenly made Coke Attorney-General.

The blow was a grave one—to Bacon, to Essex, and to the whole party; the influence of the Cecils had been directly challenged, and they had won. There was apparently a limit to the favour of the Earl. So far, however, as Bacon was concerned, a possibility still remained of retrieving the situation. Coke's appointment left the Solicitor-Generalship vacant, and it seemed obvious that Bacon was the man for the post. The Cecils themselves acquiesced; Essex felt that this time there could be no doubt about the matter; he hurried off to the Queen—and was again met by a repulse. Her Majesty was extremely reserved; she was, she said, against Bacon—for the singular reason that the only persons who supported him were Essex and Burghley. Upon that, Essex argued and expatiated, until Elizabeth lost her temper. "In passion"—so Essex told his friend in a letter written immediately afterwards—"she bade me go to bed, if I would talk of nothing else. Wherefore in passion I went away, saying while I was with her I could not but solicit for the cause and the man I so much affected, and therefore I would retire myself till I might be more graciously heard. And so we parted." And so began another strange struggle over the fate of Francis Bacon. For almost a year Elizabeth had refused to appoint an Attorney-General; was it conceivable that she was now about to delay as long in her choice of a Solicitor-General? Was it possible that, with a repetition *da capo* of all her previous waverings, she would continue indefinitely to keep every one about her in this agonising suspense?

It was, indeed, all too possible. The Solicitor-Generalship remained vacant for more than eighteen months. During all that time Essex never lost courage. He bombarded the Queen, in and out of season. He wrote to the Lord Keeper Puckering, pressing Bacon's claims; he even wrote to Sir Robert Cecil, to the same purpose. "To you, as to a Councillor," he told the latter, "I write this, that Her

Majesty never in her reign had so able and proper an instrument to do her honourable and great services as she hath now, if she will use him." Old Anthony Standen was amazed by the Earl's persistency. He had thought that his patron lacked tenacity of purpose—that "he must continually be pulled by the ear, as a boy, that learneth *ut*, *re*, *mi*, *fa*"; and now he saw that, without prompting, he was capable of the utmost pertinacity. On the other hand, in the opinion of old Lady Bacon, fuming at Gorhambury, "the Earl marred all by violent courses." The Queen, she thought, was driven to underrate the value of Francis through a spirit of sheer contradiction. Perhaps it was so; but who could prescribe the right method of persuading Elizabeth? More than once she seemed to be on the point of agreeing with her favourite. Fulke Greville had an audience of her, and, when he took the opportunity of putting in a word for his friend, she was "very exceeding gracious." Greville developed the theme of Bacon's merits. "Yes," said Her Majesty, "he begins to frame very well." The expression was perhaps an odd one; was it not used of the breaking-in of refractory horses? But Greville, overcome by the benignity of the royal manner, had little doubt that all was well. "I will lay £100 to £50," he wrote to Francis, "that you shall be her Solicitor."

While his friends were full of hope and energy,

Francis himself had become a prey to nervous agitation. The prolonged strain was too much for his sensitive nature, and, as the months dragged on without any decision, he came near to despair. His brother and his mother, similarly tempered, expressed their perturbation in different ways. While Anthony sought to drown his feelings under a sea of correspondence, old Lady Bacon gave vent to fits of arbitrary fury which made life a burden to all about her. A servant of Anthony's, staying at Gorhambury, sent his master a sad story of a greyhound bitch. He had brought the animal to the house, and "as soon as my Lady did see her, she sent me word she should be hanged." The man temporised, but "by-and-by she sent me word that if I did not make her away she should not sleep in her bed; so indeed I hung her up." The result was unexpected. "She was very angry, and said I was fransey, and bade me go home to my master and make him a fool, I should make none of her. . . . My Lady do not speak to me as yet. I will give none offence to make her angry; but nobody can please her long together." The perplexed fellow, however, was cheered by one consideration. "The bitch," he added, "was good for nothing, else I would not a hung her." The dowager, in her calmer moments, tried to turn her mind, and the minds of her sons, away from the things of this world. "I am sorry," she wrote to Anthony, "your brother with inward secret grief hindereth his health. Everybody saith he looketh thin and pale. Let him look to God, and confer with Him in godly exercises of hearing and reading, and contemn to be noted to take care."

But the advice did not appeal to Francis; he preferred to look in other directions. He sent a rich jewel to the Queen, who refused it—though graciously. He let Her Majesty know that he thought of travelling abroad; and she forbade the project, with considerable asperity. His nerves, fretted to ribbons, drove him at last to acts of indiscretion and downright folly. He despatched a letter of fiery remonstrance to the Lord Keeper Puckering, who, he believed, had deserted his cause; and he attacked his cousin Robert in a style suggestive of a female cat. "I do assure you, Sir, that by a wise friend of mine, and not factious toward your Honour, I was told with asserveration that your Honour was bought by Mr. Coventry for two thousand angels. . . . And he said further that from your servants, from your Lady, from some counsellors that have observed you in my business, he knew you wrought underhand against me. The truth of which tale I do not believe." The appointment was still hanging in the balance; and it fell to the rash and impetuous Essex to undo, with smooth words and diplomatic explanations, the damage that the wise and subtle Bacon had done to his own cause.

In October, 1595, Mr. Fleming was appointed, and

the long struggle of two and a half years was over. Essex had failed—failed doubly—failed where he could hardly have believed that failure was possible. The loss to his own prestige was serious; but he was a gallant nobleman, and his first thought was for the friend whom he had fed with hope, and whom, perhaps, he had served ill through over-confidence or lack of judgment. As soon as the appointment was made, he paid a visit to Francis Bacon. "Master Bacon," he said, "the Queen hath denied me yon place for you, and hath placed another. I know you are the least part in your own matter, but you fare ill because you have chosen me for your mean and dependence; you have spent your time and thoughts in my matters. I die if I do not somewhat towards your fortune: you shall not deny to accept a piece of land which I will bestow upon you." Bacon demurred; but he soon accepted; and the Earl presented him with a property which he afterwards sold for £1800, or at least £10,000 of our money.

Perhaps, on the whole, he had come fortunately out of the business. Worse might have befallen him. In that happy-go-lucky world, a capricious fillip from a royal finger might at any moment send one's whole existence flying into smithereens. Below the surface of caracoling courtiers and high policies there was cruelty, corruption, and gnashing of teeth. One was lucky, at any rate, not to be Mr. Booth, one of

Anthony Bacon's dependants, who, poor man, had suddenly found himself condemned by the Court of Chancery to a heavy fine, to imprisonment, and to have his ears cut off. Nobody believed that he deserved such a sentence, but there were several persons who had decided to make what they could out of it, and we catch a glimpse, in Anthony's correspondence, of this small, sordid, ridiculous intrigue, going along contemporaneously with the heroic battle over the great Law Offices. Lady Edmondes, a lady-in-waiting, had been approached by Mr. Booth's friends and offered £100 if she would get him off. She immediately went to the Queen, who was all affability. Unfortunately, however, as her Majesty explained, she had already promised Mr. Booth's fine to the head man in her stables—"a very old servant"—so nothing could be done on that score. "I mean," said her Majesty, "to punish this fool some way, and I shall keep him in prison. Nevertheless," she added, in a sudden access of generosity towards Lady Edmondes, "if your ladyship can make any good commodity of this suit, I will at your request give him releasement. As for the man's ears . . ." Her Majesty shrugged her shoulders, and the conversation ended. Lady Edmondes had no doubt that she could make a "good commodity," and raised her price to £200. She even threatened to make matters worse instead of better, as she had influence, so she declared, not only with the Queen but with the Lord Keeper Puckering. Anthony Standen considered her a dangerous woman and advised that she should be offered £150 as a compromise. The negotiation was long and complicated; but it seems to have been agreed at last that the fine must be paid, but that, on the payment of £150 to Lady Edmondes, the imprisonment would be remitted. Then there is darkness; in low things as in high the ambiguous Age remains true to its character; and, while we search in vain to solve the mystery of great men's souls and the strange desires of Princes, the fate of Mr. Booth's ears also remains for ever concealed from us.

VI

MR. BOOTH'S case was a brutal farce, and the splendid Earl, busied with very different preoccupations—his position with the Queen, the Attorney-Generalship, the foreign policy of England—could hardly have given a moment's thought to it. But there was another criminal affair no less obscure but of far more dreadful import which, suddenly leaping into an extraordinary notoriety, absorbed the whole of his attention—the hideous tragedy of Dr. Lopez.

Ruy Lopez was a Portuguese Jew who, driven from the country of his birth by the Inquisition, had come to England at the beginning of Elizabeth's reign and set up as a doctor in London. He had been extremely successful; had become house physician at St. Bartholomew's Hospital; had obtained, in spite of professional jealousy and racial prejudice, a large practice among persons of distinction; Leicester and Walsingham were his patients; and, after he had been in England for seventeen years, he reached the highest place in his profession: he was made physician-in-chief to the Queen. It was only natural that there should have been murmurs against a Jewish foreigner who had outdone his English rivals; it was rumoured that he owed his advancement less to medical skill than flattery and self-advertisement; and in a libellous pamphlet against Leicester it was hinted that he had served that nobleman all too well—by distilling his poisons for him. But Dr. Lopez was safe in the Queen's favour, and such malice could be ignored. In October, 1593, he was a prosperous elderly man—a practising Christian, with a son at Winchester, a house in Holborn, and all the appearances of wealth and consideration.

His countryman, Don Antonio, the pretender to the Portuguese crown, was also living in England. Since the disastrous expedition to Lisbon four years earlier, this unfortunate man had been rapidly sinking into disrepute and poverty. The false hopes which he had held out of a popular rising on his behalf in Portugal had discredited him with Elizabeth. The magnificent jewels which he had brought with him to England had been sold one by one; he was surrounded by a group of famishing attendants; fobbed off with a meagre pension, he was sent, with his son, Don Manoel, to lodge in Eton College, whence, when the Queen was at Windsor, he would issue forth, a haggard spectre, to haunt the precincts of the Court.

Yet he was still not altogether negligible. He still might be useful as a pawn in the game against Spain. Essex kept a friendly eye upon him, for the Earl, by an inevitable propulsion, had become the leader of the anti-Spanish party in England. The Cecils, naturally pacific, were now beginning to hope that the war, which seemed to be dragging on by virtue rather of its own impetus than of any good that it could do to either party, might soon be brought to an end. This was enough in itself to make Essex bellicose; but he was swayed not merely by opposition to the Cecils; his restless and romantic temperament urged him irresistibly to the great adventure of war; thus only could his true nature express itself, thus only could he achieve the glory he desired. Enemies he must have: at home—who could doubt it?—the Cecils; abroad—it was obvious—Spain! And so he became the focus of the new Elizabethan patriotism—a patriotism that was something distinct from religion or policy—that was the manifestation of that enormous daring, that superb self-confidence, that thrilling sense of solidarity, which, after so many years of doubt and preparation, had come to the English race when the smoke had rolled away and the storm subsided, and there was revealed the wreck of the Armada. The new spirit was resounding, at that very moment, in the glorious rhythm of Tamburlaine; and its living embodiment was Essex. He would assert the greatness of England in unmistakable fashion—by shattering the power of the Spaniard once for all. And in such an enterprise no instrument must be neglected; even the forlorn Don Antonio might prove serviceable yet. There might—who knew?—be another expedition to Portugal, more fortunate than the last. King Philip, at any rate, was of that opinion. He was extremely anxious to get Don Antonio out of the way. More than one plot for his assassination had been hatched at Brussels and the Escurial. His needy followers, bought by Spanish gold, crept backwards and forwards between England and Flanders, full of mischief. Anthony Bacon, through his spies, kept a sharp look-out. The pretender must be protected; for long he could lay his hands on nothing definite; but one day his care was rewarded.

News reached Essex House that a certain Esteban Ferreira, a Portuguese gentleman, who had been ruined by his adherence to the cause of Don Antonio, and was then living in Lopez's house in Holborn, was conspiring against his master and had offered his services to the King of Spain. The information was certainly trustworthy, and Essex obtained from Elizabeth an order for the arrest of Ferreira. The man was accordingly seized; no definite charge was brought against him, but he was put into the custody of Don Antonio at Eton. At the same time instructions were sent to Rye, Sandwich and Dover, ordering all Portuguese correspondence that might arrive at those ports to be detained and read. When Dr. Lopez heard of the arrest of Ferreira, he went to the Queen and begged for the release of his countryman. Don Antonio, he said, was much to blame; he treated his servants badly; he was ungrateful to Her Majesty. Elizabeth listened, and the Doctor ventured to observe that Ferreira, if released, might well be employed to "work a peace between the two kingdoms." This suggestion seemed not to please Elizabeth. "Or," said the Doctor, "if your Majesty does not desire that course . . ." he paused, and then added, enigmatically, "might not a deceiver be deceived?" Elizabeth stared; she did not know what the fellow meant, but he was clearly taking a liberty. She "uttered"—so we are told by Bacon—"dislike and disallowance"; and the Doctor, perceiving that he had not made a good impression, bowed himself out of the room.

A fortnight later, Gomez d'Avila, a Portuguese of low birth, who lived near Lopez's house in Holborn, was arrested at Sandwich. He was returning from Flanders, and a Portuguese letter was discovered upon his person. The names of the writer and the addressee were unknown to the English authorities. The contents, though they appeared to refer to a commercial transaction, were suspicious; there were phrases that wore an ambiguous look. "The bearer will inform your Worship in what price your pearls are held. I will advise your Worship presently of the uttermost penny that can be given for them. . . . Also this bearer shall tell you in what resolution we rested about a little musk and amber, the which I determined to buy. . . . But before I resolve myself I will be advised of the price thereof; and if it shall please your Worship to be my partner, I am persuaded we shall make good profit." Was there some hidden meaning in all this? Gomez d'Avila would say nothing. He was removed to London, in close custody. When there, while waiting in an antechamber before being examined by those in charge of the case, he recognised a gentleman who could speak Spanish. He begged the gentleman to take the news of his arrest to Dr. Lopez.

Meanwhile, Ferreira was still a prisoner at Eton. One day he took a step of a most incriminating kind. He managed to convey to Dr. Lopez, who had taken lodgings close by, a note, in which he warned the Doctor "for God's sake" to prevent the coming over of Gomez d'Avila from Brussels, "for if he should be taken the Doctor would be undone without remedy." Lopez had not yet heard of the arrest of Gomez, and replied, on a scrap of paper hidden in a handkerchief, that "he had already sent twice or thrice to Flanders to prevent the arrival of Gomez, and would spare no expense, if it cost him £300." Both the letters were intercepted by Government spies, read, copied and passed on. Then Ferreira was sent for, confronted with the contents of his letter, and informed that Dr. Lopez had betrayed him. He immediately declared that the Doctor had been for years in the pay of Spain. There was a plot, he said, by which Don Antonio's son and heir was to be bought over to the interests of Philip; and the Doctor was the principal agent in the negotiations. He added that, three years previously, Lopez had secured the release from prison of a Portuguese spy, named Andrada, in order that he should go to Spain and arrange for the poisoning of Don Antonio. The information was complicated and strange; the authorities took a careful note of it; and waited for further developments.

At the same time, Gomez d'Avila was shown the rack in the Tower. His courage forsook him, and he confessed that he was an intermediary, employed to carry letters backwards and forwards between Ferreira in England and another Portuguese, Tinoco, in Brussels, who was in the pay of the Spanish Government. The musk and amber letter, he said, had been written by Tinoco and addressed to Ferreira, under false names. Gomez was then plied with further questions, based upon the information obtained from Ferreira. It was quite true, he admitted, that there was a plot to buy over Don Antonio's son. The youth was to be bribed with 50,000 crowns and the musk and amber letter referred to this transaction. Ferreira, examined in his turn, confessed that this was so.

Two months later Burghley received a communication from Tinoco. He wished, he said, to go to England, to reveal to the Queen secrets of the highest importance for the safety of her realm, which he had learnt at Brussels; and he asked for a safe-conduct. A safe-conduct was despatched; it was, as Burghley afterwards remarked, "prudently drafted"; it allowed the bearer safe ingress into England, but it made no mention of his going away again. Shortly afterwards Tinoco arrived at Dover; upon which he was at once arrested, and taken to London. His person was searched, and bills of exchange for a large sum of money were found upon him, together with two letters from the Spanish governor of Flanders, addressed to Ferreira.

Tinoco was a young man who had been through much. For years he had shared the varying fortunes of Don Antonio; he had fought in Morocco, had been taken prisoner by the Moors, and after four years of slavery had rejoined his master in England. Destitute and reckless, he had at last, like his comrade Ferreira, sold himself to Spain. What else could such creatures do? They were floating straws sucked into the whirlpool of European statecraft; they had no choice; round and round they eddied, ever closer to the abyss. But for Tinoco, who was young, strong, and courageous, a life of treachery and danger had, perhaps, its attractions. There was a zest in the horror; and, besides, Fortune was capricious; the bold, unscrupulous intriguer might always pull some golden prize from the lottery, as well as some unspeakably revolting doom.

The letters found on his person were vague and mysterious, and some sinister interpretation might well be put upon them. They were sent to Essex, who decided himself to interrogate the young man. The examination was conducted in French; Tinoco had a story ready—that he had come to England to reveal to the Queen a Jesuit plot against her life; but he broke down under the cross-examination of the Earl, prevaricated, and contradicted himself. Next day he wrote a letter to Burghley, protesting his innocence. He had been, he said, "confused and encumbered by the cunning questions of the Earl of Essex"; with his small knowledge of French, he had failed to understand the drift of the inquiry, or to express his own meaning; and he begged to be sent back to Flanders. The only result of his letter was that he was more rigorously confined. Again examined by Essex, and pressed with leading questions, he avowed that he had been sent to England by the Spanish authorities in order to see Ferreira and with him to win over Dr. Lopez to do a service to the King of Spain. Dr. Lopez once more! Every line of enquiry, so it seemed to Essex, led straight to the Jew. His secret note to Ferreira had been deeply incriminating. Ferreira himself, Gomez d'Avila, and now Tinoco all agreed that the Doctor was the central point in a Spanish conspiracy. That conspiracy, if they were to be believed, was aimed against Don Antonio; but could they be believed? Might not some darker purpose lie behind? The matter must be sifted to the bottom. Essex went to the Queen; and on the 1st January, 1594, Dr. Lopez, principal physician to Her Majesty, was arrested.

He was taken to Essex House, and there kept in close custody, while his house in Holborn was searched from top to bottom; but nothing suspicious was found there. The Doctor was then examined by the Lord Treasurer, Robert Cecil, and Essex. He had a satisfactory answer for every question. The Cecils were convinced that Essex had discovered a mare's nest. In their opinion, the whole affair was merely a symptom of the Earl's anti-Spanish obsession; he saw plots and spies everywhere; and now he was trying to get up a ridiculous agitation against this unfortunate Jew, who had served the Queen faithfully for years, who had furnished an explanation of every suspicious circumstance, and whose general respectability was a sufficient guarantee that this attack on him was the result of folly and malice. Accordingly, as soon as the examination was over, Sir Robert hurried to the Queen, and informed her that both his father and himself were convinced of the Doctor's innocence. But Essex was still unshaken; he persisted in the contrary opinion. He too went to the Queen, but he found her with Sir Robert, and in a passion. As soon as he appeared, he was overwhelmed with royal invectives. Elizabeth declared that he was "a rash and temerarious youth," that he had brought accusations against the Doctor which he could not prove, that she knew very well the poor man was innocent, that she was much displeased, and that her honour was at stake in the matter. The flood of words poured on, while Essex stood in furious silence, and Sir Robert surveyed the scene with gentle satisfaction. At last the Earl, his expostulations cut short with a peremptory gesture, was dismissed from the presence. He immediately left the palace, hurried to his house and, brushing aside his attendants without a word or a look, shut himself into his room and flung himself upon his bed in an agony of wrath and humiliation. For two days he remained there, silent and enraged. At length he emerged, with fixed determination in his countenance. *His* honour, no less than the Queen's was at stake; come what might, he must prove the Cecils to be utterly mistaken; he must bring Dr. Lopez to justice.

Characteristically enough, in spite of the Queen's anger and the Cecils' scepticism, the case against the Doctor was not allowed to drop. He was still kept a prisoner at Essex House; he and the rest of the suspected Portuguese were still subjected to endless examinations. And now began one of those strange and odious processes which fill the obscure annals of the past with the ironical futility of human justice. The true principles of criminal jurisprudence have only come to be recognised, with gradually increasing completeness, during the last two centuries; the comprehension of them has grown with the growth of science—with the understanding of the nature of evidence, and the slow triumph, in men's mental habits, of ordered experience and reason. No human creature can ever hope to be truly just; but there are degrees in mortal fallibility, and for countless ages the justice of mankind was the sport of fear, folly, and superstition. In the England of Elizabeth there was a particular influence at work which, in certain crucial cases, turned the administration of justice into a mockery. It was virtually impossible for any one accused of High Treason—the gravest offence known to the law—to be acquitted. The reason for this was plain; but it was a reason, not of justice, but expediency. Upon the life of Elizabeth hung the whole structure of the State. During the first thirty years of her reign, her death would have involved the accession of a Catholic sovereign, which would inevitably have been followed by a complete revolution in the system of Government, together with the death or ruin of the actual holders of power. The fact was obvious enough to the enemies of the

English polity, and the danger that they might achieve their end by the Queen's assassination was a very real one. The murder of inconvenient monarchs was one of the habits of the day. William of Orange and Henry III of France had both been successfully obliterated by Philip and the Catholics. Elizabeth on her side had sought—though, indeed, rather half-heartedly—to have the Queen of Scots secretly put out of the way, in order to avoid the public obloquy of a judicial execution. Her own personal fearlessness added to the peril. She refused, she said, to mistrust the love of her subjects; she was singularly free of access; and she appeared in public with a totally inadequate guard. In such a situation, only one course of action seemed to be possible: every other consideration must be subordinated to the supreme necessity of preserving the Queen's life. It was futile to talk of justice; for justice involves, by its very nature, uncertainty; and the government could take no risks. The old saw was reversed; it was better that ten innocent men should suffer than that one guilty man should escape. To arouse suspicion became in itself a crime. The proofs of guilt must not be sifted by the slow processes of logic and fair play; they must be multiplied—by spies, by *agents provocateurs*, by torture. The prisoner brought to trial should be allowed no counsel to aid him against the severity of iron-hearted judges and the virulence of the ablest lawyers of the day. Conviction should be followed by the most frightful of punishments. In the domain of treason, under Elizabeth, the reign of law was, in effect, superseded, and its place was taken by a reign of terror.

It was in the collection of evidence that the mingled atrocity and absurdity of the system became most obvious. Not only was the fabric of a case often built up on the allegations of the hired creatures of the government, but the existence of the rack gave a preposterous twist to the words of every witness. Torture was constantly used; but whether, in any particular instance, it was used or not, the consequences were identical. The threat of it, the hint of it, the mere knowledge in the mind of a witness that it might at any moment be applied to him—those were differences merely of degree; always, the fatal compulsion was there, inextricably confusing truth and falsehood. What shred of credibility could adhere to testimony obtained in such circumstances—from a man, in prison, alone, suddenly confronted by a group of hostile and skilful examiners, plied with leading questions, and terrified by the imminent possibility of extreme physical pain? Who could disentangle among his statements the parts of veracity and fear, the desire to placate his questioners, the instinct to incriminate others, the impulse to avoid, by some random affirmation, the dislocation of an arm or a leg? Only one thing was plain about such evidence:—it would always be possible to give to it whatever interpretation the prosecutors might desire. The Government could prove anything. It could fasten guilt upon ten innocent men with the greatest facility. And it did so, since by no other means could it make certain that the one actual criminal—who might be among them—should not escape. Thus it was that Elizabeth lived her life out, unscathed; and thus it happened that the glories of her Age could never have existed without the spies of Walsingham, the damp cells of the Tower, and the notes of answers, calmly written down by cunning questioners, between screams of agony.

It was, of course, an essential feature of the system that those who worked it should not have realised its implications. Torture was regarded as an unpleasant necessity; evidence obtained under it might possibly, in certain cases, be considered of dubious value; but no one dreamt that the judicial procedure of which it formed a part was necessarily without any value at all. The wisest and the ablest of those days—a Bacon, a Walsingham—were utterly unable to perceive that the conclusions, which the evidence they had collected seemed to force upon them, were in reality simply the result of the machinery they themselves had set in motion. Judges, as well as prisoners, were victims of the rack.

The case of Dr. Lopez was typical. One can trace in it the process by which suspicion, fear, and preconceived theories were gradually, under the pressure of the judicial system, blended into a certainty which, in fact, was baseless. Essex was an honest young lord, who would have recoiled in horror from the thought of doing an innocent man to death for political purposes; but he was not very strong in the head. He mistrusted the Cecils, he mistrusted Spain, he perceived—what was true enough—that there was something fishy about Dr. Lopez. The scorn poured by the Queen upon his sagacity was the final inducement: he was right, in spite of them all; he would not rest till he had probed the matter to the bottom. And there was only one method of effecting this—it was obvious; the Portuguese must be cross-examined until the truth was forced from them. Lopez himself had baffled him, but there remained Ferreira and Tinoco, who had already shown themselves more pliable. They were accordingly, in their separate cells, relentlessly questioned. Each was ready enough, in order to exculpate himself, to incriminate the other, and to declare, when pressed further, that the Doctor was the centre of the plot. But what was the plot? If it was merely aimed at Don Antonio, why this elaboration of mystery? But if it was aimed at some one else? If . . . ? It needed no genius to unravel the enigma. One had only to state the circumstances, for the solution to arise spontaneously to the mind. Spain—a plot—the royal physician: such a concatenation was enough. It was one more attempt on the part of King Philip to assassinate the Queen of England.

This point once reached, the next step inevitably followed. The belief in the mind of the questioner became a statement in the mouth of the questioned. At one point in his examination, Ferreira asserted that Dr. Lopez had written to the King of Spain, professing his willingness to do everything his Majesty required. The question was then asked—"Would the Doctor have poisoned the Queen if required?" and Ferreira replied in the affirmative. He was then forced to elaborate the supposition with a mass of detail; and the same process was applied to Tinoco; with the same result. After that, supposition very soon slipped into fact. "I have discovered," wrote Essex to Anthony Bacon, "a most dangerous and desperate treason. The point of conspiracy was her Majesty's death. The executioner should have been Dr. Lopez; the manner poison. This I have so followed that I will make it appear as clear as noonday."

Luck was against the Doctor. The case against him depended on a complicated construction from the evidence of two perjured rogues, Ferreira and Tinoco—evidence extorted under fear of the rack, and made up of a mass of hearsay and the recollections of years' old conversations and of letters never produced. The

Cecils, with their pro-Spanish and anti-Essex bias, would have been sharp enough to see through such stuff, but for one unfortunate circumstance. Early in the proceedings the name of Andrada, a Portuguese spy, had been mentioned by Ferreira, who had asserted that he had been sent to Spain by Lopez to arrange for the murder of Don Antonio. Andrada was well known to Burghley. It was true that the man had been to Spain, at the period mentioned, in most suspicious circumstances. Burghley had no doubt that, while nominally in the service of Don Antonio, he had been bought by the Spanish authorities. He was now in Brussels; and, if it was a fact that there had been a secret connection between him and Lopez, something really damaging would at last have been discovered about the Doctor. As the examinations proceeded, Andrada's name recurred more and more frequently. It appeared that he had been the principal intermediary between the Spanish Court and the intriguers in Flanders. Tinoco repeated—or purported to repeat—a long description that Andrada had given of his visit to Madrid. King Philip had embraced him, and told him to pass on the embrace to Dr. Lopez; he had handed him a diamond and ruby ring, with a similar injunction. Could all this be true? Elizabeth was told of it, and she remembered that, some three years previously, the Doctor had offered her a diamond and ruby ring, which she had refused to accept. The Doctor was now pressed once more with searching questions. He denied, with violent oaths and imprecations, that he knew anything of the matter; but at last, when cross-examined on the ring, he changed his tone. It was true, he admitted, that he had been privy to Andrada's visit to Spain; but he added that the explanation of that visit was entirely different from any that had been put forward. Andrada had been in the pay of Walsingham. He had been sent to Madrid on the pretext of a peace negotiation, with the object of spying out the state of affairs at the Spanish Court. The Doctor, at Walsingham's special request, had agreed to allow his name to be used, to give colour to the proceedings. Andrada was to represent to Philip that he had been sent by Lopez, who was eager for peace and influential with the Queen. The deceiver, in fact, was to be deceived. The scheme had worked, Philip had been taken in, and his ring had been intended, not for the Doctor, but for Elizabeth. Walsingham was perfectly aware of all this, and could substantiate every detail. Could, that is to say, if only . . . Essex laughed outright. The Cecils, convinced that Andrada was in the pay of the Spaniards, were incredulous. It would not do. The Doctor's story was ingenious—it was too ingenious; the whole—it was obvious—hung upon one thing—the corroboration of Walsingham; and Walsingham was dead.

By a curious irony, the very circumstance which finally led the Cecils to abandon Lopez has afforded to posterity the means of vindicating him. Papers have been discovered among the Spanish archives showing that his tale was substantially true. It was indeed under the pretext of a peace overture that Andrada visited Madrid. He was not permitted to see Philip in person, and the story of the royal embrace was a fabrication; but the diamond and ruby ring was actually handed to the spy by the Spanish Secretary of State. Other matters, it is true, were discussed besides peace; it was agreed that Dr. Lopez should endeavour to obtain either the imprisonment of Don Antonio, or his exile from England; a hint was thrown out that he might usefully be poisoned; but not the faintest suggestion was made which could possibly point to the murder of Elizabeth. As a matter of fact, however, and this was unknown to Lopez—the Spaniards were not taken in. They saw through Walsingham's stratagem, and they determined to hoist him with his own petard. Persuaded by their gold, Andrada became a double spy. He agreed to return to England and to carry on, nominally, the negotiation for peace, but, in reality, to use his position for furnishing Madrid with inside information of the state of affairs in England. Walsingham's death spoilt the plan. Andrada was unable to explain his conduct, and Burghley became convinced

A form of trial followed. Ferreira and Tinoco, far from saving themselves by their incriminations of the Doctor, were arraigned beside him as accomplices in his guilt. Tinoco in vain pleaded the protection of his safe-conduct; the lawyers solemnly debated the point, and decided against him. All three were sentenced to the death of traitors. The popular excitement was intense. As Essex had foreseen, the hatred of Spain, which had been dying down, rose again to a frenzy throughout the country. Dr. Lopez became the type of the foreign traitor, and his villainy was sung in ballads, and his name hissed with execrations from the boards of theatres. That he was a Jew was merely an incidental iniquity, making a shade darker the central abomination of Spanish intrigue. Modern critics have seen in him the original of Shylock, who appeared upon the stage a few years later; but such a supposition is wide of the mark. In fact, if Shakespeare thought of Dr. Lopez at all in connection with Shylock, it must have been because of his unlikeness, and not of his resemblance, to the great figure in "The Merchant of Venice." The two characters are antithetical. The whole essence of Shylock lies in his colossal, his tragic, Hebraism; but Dr. Lopez was Europeanised and Christianised—a meagre, pathetic creature, who came to his ruin by no means owing to his opposition to his gentile surroundings, but because he had allowed himself to be fatally entangled in them. Yet, perhaps, it is not fanciful to imagine that Shakespeare, in his tragedy of the Venetian outcast, glanced for a moment, under cover of a piece of amorous jesting, at that other tragedy of the royal physician. "Ay," says Portia to Bassanio,

> "but I fear you speak upon the rack,
> Where men enforcèd do speak anything."[1]

The wisdom and the pity of the divine poet exquisitely reveal themselves in those light words.

The Queen hesitated even more than usual before she allowed the sentences to be carried into execution. Possibly she was waiting for some confirmation or some denial from the authorities in Spain or Flanders; possibly, in spite of all the accumulated proof of the Doctor's guilt, she was unable to obliterate from her mind her instinctive perception of his innocence. Four months elapsed before she allowed the law to take its course. Then—it was June, 1594—the three men, bound to hurdles, were dragged up Holborn, past the Doctor's house, to Tyburn. A vast crowd was assembled to enjoy the spectacle. The Doctor standing on the scaffold attempted in vain to make a dying speech; the mob was too angry and too delighted to be quiet; it howled with laughter, when, amid the uproar, the Jew was heard asseverating that he loved his mistress better than Jesus Christ; no more was heard, and the old man was hurried to the gallows. He was strung up and—such was the routine of the law—cut down while life was still in him. Then the rest of the time-honoured punishment—castration, disembowelling, and quartering—was carried out. Ferreira was the next to suffer. After that, it was the turn of Tinoco. He had seen what was to be his fate, twice repeated, and from close enough. His ears were filled with the shrieks and the moans of his companions, and his eyes with every detail of the contortions and the blood. And so his adventures had ended thus at last. And yet, they had not quite ended; for Tinoco, cut down too soon, recovered his feet after the hanging. He was lusty and desperate; and he fell upon his executioner. The crowd, wild with excitement, and cheering on the plucky foreigner, broke through the guards, and made a ring to watch the fight. But, before long, the instincts of law and order reasserted themselves. Two stalwart fellows, seeing that the executioner was giving ground, rushed forward to his rescue. Tinoco was felled by a blow on the head; he was held down firmly on the scaffold; and, like the others, castrated, disembowelled, and quartered.

Elizabeth was merciful to the Doctor's widow. She allowed her to keep the goods and chattels of the deceased, forfeited by his attainder—with one exception. She took possession of King Philip's ring. She slipped it—who knows with what ironical commiseration?—on to her finger; and there it stayed till her death.

QUEEN ELIZABETH IN 1596

From a contemporary engraved portrait

VII

THE Spanish question grew ever more acute. A war that was no war might exactly suit the temper of Elizabeth; but it seemed an infamy to Essex, and was no less distasteful to Henry of France, pressed hard by the Spaniards on his northern frontier and by the Catholic Leaguers in his own dominions. The French king and the English peer came together in a curious combination. Their joint object was to propel Elizabeth into an alliance with France, which would involve the active participation of England in an attack on the Spaniards. Between them flew, backwards and forwards, uniting and enflaming their energies, the stormy petrel, Antonio Perez, in whom a frantic hatred of King Philip had become the very breath of life.

A few years earlier Perez had fled from Spain in the wildest circumstances. Philip's principal Secretary of State, he had quarrelled with his master over a murder, had taken refuge in his native town of Saragossa and had there, at the King's instigation, been seized by the Inquisition. His fate seemed certain; but unexpected forces came to his rescue, and Perez lives in history as the one man, who, having once fallen into the clutches of the Holy Office, escaped with a whole skin. The charges against him were, indeed, highly serious. Exasperated in a dungeon, the misguided secretary had allowed himself, in his ravings, to insult not only the King but the Deity. "God sleeps! God sleeps!" he had exclaimed, and his words had been heard and noted. "This proposition," the official report declared, "is heretical, as if God had no care for human beings, when the Bible and the Church affirm that He does care." That was bad enough, but worse followed. "If it is God the Father," said the miscreant, "who has allowed the King to behave so disloyally towards me, I'll pull God the Father's nose!" "This proposition," said the official report, "is blasphemous, scandalous, offensive to pious ears, and savouring of the heresy of the Vaudois, who affirmed that God was corporeal and had human members. Nor is it an excuse to say that Christ, being made man, had a nose, since the words were spoken of the First Person of the Trinity." The stake was the obvious retribution for such wickedness, and the proper preparations were being made when the people of Saragossa suddenly rose in arms. The ancient liberties of Aragon, its immemorial rights of jurisdiction, were being infringed, they asserted, by the King and the Holy Office. They invaded the prison, beat to death the royal governor, and set Perez free. He escaped to France; but his safety proved expensive to Saragossa. For soon afterwards the King's army appeared upon the scene, and the ancient liberties of Aragon were finally abolished, while seventy-nine of the popular party were burnt alive in the market-place, the ceremony beginning at eight in the morning and ending at nine in the evening, by torchlight.

The hectic hero of this affair was now leading the life of an exile and an intriguer. He was obviously a rogue, but he might, for the moment at any rate, be a useful rogue; and on that footing he had won his way into the good graces of Essex and Henry. He was active and unscrupulous; he was full of stories that were infinitely discreditable to the King of Spain, and he was master of an epistolary style of Euphuistic Latin which precisely hit off the taste of the great ones of that generation. How delightful to weave plots, change policies, and direct the fate of Europe in learned antitheses and elegant classical allusions!

When the conclave at Essex House judged that the time was ripe, a letter was despatched from the Earl to Perez, hinting that, if Henry really wished for Elizabeth's alliance, his best course was to threaten to make peace himself with Spain. If Juno was France and Philip the King of the Underworld, was not the conclusion clear? For who was so ignorant as not to know that Juno, when she had implored for help many times and in vain, had at last burst out with—

"Flectere si nequeo superos, Acheronta movebo"? "But silence, my pen! And silence Antonio! For methinks I have read the poets too much." [1]

Perez at once showed the letter to Henry, who was not slow to catch its drift. Taking the advice of his English friend, he despatched a special envoy to Elizabeth, with instructions to inform her that he had received favourable offers of peace from Spain, and was inclined to accept them. Elizabeth was apparently unmoved by this intelligence; she wrote a letter of expostulation to Henry, but she was unable, she declared, to give him further help; yet she was secretly uneasy, and soon afterwards despatched, on her side, a special envoy to France, who was to discover and report to her the real inclinations of the King.

This envoy was Sir Henry Unton, one of those remarkable ambassadors who divided their allegiance between the Government and Essex House. He went to France armed with the instructions, not only of Elizabeth, but of Anthony Bacon. A letter exists in which Unton is directed, with minute detail, to inform the French King that he must hold firm; in which he is told so to arrange matters as to be received with public coldness by Henry; and to "send us thundering letters, whereby he must drive us to propound and

[1] Juno autem, quum saepius frustra spem implorasset, tandem eripuit: "Flectere si nequeo superos, Acheronta movebo." . . . Sed tace, calame, et tace, Antoni, nimium enim poetas legisse videor.

to offer." Unton did as he was bid, and the thundering letters duly arrived. At the same time, Perez had been ordered to write to the Earl "such a letter as may be showed, wherein he shall say that the sending of Unton hath made all things worse than ever." Perez too was all obedience; he sent off, in elaborate Latin, a report of Henry's asseverations in favour of peace; he himself, he added, could not understand the policy of the English Government; but perhaps there was some mystery that was unrevealed—"the designs of Princes are a deep abyss." [1]

It was perfectly true. All the letters were shown to the Queen, who read them carefully through, with a particular relish for the latinity of Perez. But the result of this extraordinary intrigue was not at all what might have been expected. Perhaps Elizabeth had smelt a rat. However that may be, she calmly wrote to Henry that she was very ready to help him against Spain with men and money—on one condition: that he should give into her keeping the town of Calais. The charming proposal was not well received. "I had as lief be bitten by a dog as scratched by a cat," exclaimed the infuriated Béarnais. But in a few weeks he found that he had spoken more truly than he thought. A Spanish army advanced from Flanders, laid siege to Calais, and stormed the outworks of the town. The roar of the besieging guns could be dis-

[1] Fines principum abyssus multa.

tinctly heard—so Camden tells us—in the royal palace at Greenwich.

Elizabeth did not like that. Not only was the noise disturbing, but the presence of the Spaniards in a port commanding the narrow seas would be distinctly inconvenient. The next news was that the town of Calais had fallen, but that the citadel still held out. Something might yet be done, and a hasty levy of men was raised in London, and sent down with all speed, under the command of Essex, to Dover. With luck, the French might be relieved and the situation saved; but it suddenly occurred to Elizabeth that, with luck also, the French might relieve themselves, and that in any case the whole thing was too expensive. Accordingly, when the troops were actually on board, a courier galloped down to the shore with a letter from the Queen countermanding the expedition. Essex raved and implored with his usual energy; but, while the messengers were posting to and fro between Dover and London, the Spaniards took the citadel (April 14th, 1596).

This was too much, even for the hesitancy of Elizabeth. She could not conceal from herself that, in this instance, at any rate, she had failed; that the beautiful negation, which was the grand object of all her policy, had eluded her; that, in fact, something had actually occurred. She was very angry, but the necessity for some sort of action on her own part gradually forced itself upon her; and for the first time she began to listen seriously to the suggestions of the war party.

There were two possibilities of attack. A really effective army might be sent to France which would be strong enough to enable Henry to deal with the Spaniards. This was the course that Perez, accompanied by the Duc de Bouillon, was immediately despatched across the Channel to urge, with all the fury of his eloquence, upon Elizabeth. But when the emissaries arrived they found to their astonishment that the wind had changed in England. Another project was on foot. For months a rebellion had been simmering in Ireland, and there was reason to believe that Philip was busy fitting out an expedition to give succour to his Catholic friends. It was now proposed to forestall his offensive by delivering a naval attack upon Spain. Essex was suddenly converted to the plan. Throwing over Henry and Perez with gay insouciance, he pressed upon the Queen the formation of a powerful armament to be sent not to Calais, but to Cadiz. Elizabeth consented. She appointed Essex and the Lord Admiral Howard of Effingham joint commanders of the force; and, within a fortnight of the fall of Calais, the Earl was in Plymouth, collecting together in feverish energy an army and a fleet.

Elizabeth had consented; but, in the absence of Essex, the suggestions of Perez sounded sweetly in her ear. She began wavering once again. Perhaps, after all, it would be wiser to help the French King; and surely it would be dangerous to send off the fleet on a wild-cat expedition—the fleet, which was her one protection against a Spanish invasion. The news of her waverings reached Essex, and filled him with agitation. He knew too well the temper of his mistress. "The Queen," he wrote, "wrangles with our action for no cause but because it is in hand. If this force were going to France, she would then fear as much the issue there as she doth our intended journey. I know I shall never do her service but against her will." He had racked his wits, he added, to bring her to agree to the expedition, and if it fell through now he swore he would "become a monk upon an hour's warning."

Certainly, it was touch and go. The next news was that an offensive and defensive league had been concluded with France; and a few days later the Queen wrote a letter to the two Lord Generals at Plymouth, which seemed to portend yet another change of policy. They were ordered to put the expedition under the command of some inferior officers, and to return themselves to the royal presence—"they being so dear unto her and such persons of note, as she could not allow of their going." The Court was in a ferment. As the terrible moment of decision approached, Elizabeth's mind span round like a teetotum. She was filled with exasperation and rage. She thundered against Essex, who, she said, was forcing her to do this thing against her will. The oldest courtiers were appalled, and Burghley, with trembling arguments and venerable aphorisms, sought in vain to appease her. The situation was complicated by the reappearance of Walter Raleigh. He had returned from Guiana, more exuberant and formidable than ever, with endless tales of wealth and adventure, and had been received with something like forgiveness by the Queen. Was it possible that the recall of Essex and Howard would be followed by the appointment of Raleigh to the supreme command? But the expedition itself, even if it was sanctioned, and whoever commanded it, might never start, for the difficulties in the way of its preparation were very great, there was a shortage of men, of money, of munitions, and it almost looked as if the armament would only be ready when it was too late to be of any use. Confusion reigned; anything might happen; then, all at once, the fog rolled off, and certainty emerged. Elizabeth, as was her wont, after being buffeted for so long and in so incredible a fashion by a sea of doubts, found herself firmly planted on dry land. The expedition was to go—and immediately; Essex and Howard were reinstated, while Raleigh was given a high, though subordinate, command. The new orientation of English policy was signalised in a curious manner—by the degradation of Antonio Perez. The poor man was no longer received at Court; he took no part in the final stages of the French treaty; the Cecils would not speak to him; he sought refuge in desperation with Anthony Bacon, and Anthony Bacon was barely polite. His life of vertiginous intrigue suddenly collapsed. Back in France again he was looked upon with coldness, with faint animosity. He faded, dwindled, and sank; and when, years later, worn out with age and poverty, he expired in a Parisian garret, the Holy Office may well have felt that the sufferings of the enemy who had escaped its vengeance must have been, after all, almost enough.

In the midst of his agitations at Plymouth, Essex had received a letter from Francis Bacon. The Lord Keeper Puckering had died; Egerton, the Master of the Rolls, had been appointed to succeed him; and Bacon now hoped for Egerton's place. He wrote to ask for the Earl's good offices, and his request was immediately granted. Pressed and harassed on every side by the labours of military organisation, by doubts of the Queen's intentions, by anxieties over his own position, Essex found the time and the energy to write three letters to the leaders of the Bar, pressing upon them, with tactful earnestness, the claims of his friend. Francis was duly grateful. "This accumulating," he wrote, "of your Lordship's favours upon me hitherto worketh only this effect: that it raiseth my mind to aspire to be found worthy of them, and likewise to merit and serve you for them." But whether, he added, "I shall be able to pay my vows or no, I must leave that to God, who hath them *in deposito*."

Among all the confusions that surrounded the departure of the expedition, not the least disturbing were those caused by the antagonism of the two commanders. Essex and Lord Howard were at loggerheads. They bickered over everything from the rival claims of the army and the navy to their own places in the table of precedence. Howard was Lord Admiral, but Essex was an Earl; which was the higher? When a joint letter to the Queen was brought for their signature, Essex, snatching a pen, got in his name at the top, so that Howard was obliged to follow with his underneath. But he bided his time—until his rival's back was turned; then, with a pen-knife, he cut out the offending signature; and in that strange condition the missive reached Elizabeth.

Everything was ready at last; it was time to say farewell. The Queen, shut up in her chamber, was busy with literary composition. The results of her labour were entrusted to Fulke Greville, who rode down with the final despatches to Plymouth and handed them to Essex. There was a stately private letter from the Queen to the General:—"I make this humble bill of requests to Him that all makes and does, that with His benign hand He will shadow you so, as all harm may light beside you, and all that may be best hap to your share; that your return may make you better and me gladder." There was a friendly note from Robert Cecil, with a last gay message from Elizabeth. "The Queen says, because you are poor she sends you five shillings." And, in addition, there was a royal prayer, to be read aloud to the assembled forces, for the success of the expedition. "Most omnipotent and guider of all our world's mass! that only searchest and fathomest the bottoms of all hearts and conceits, and in them seest the true original of all actions intended . . . . Thou, that diddest inspire the mind, we humbly beseech, with bended knees, prosper the work and with best forewinds guide the journey, speed the victory, and make the return the advancement of thy fame and surety to the realm, with least loss of English blood. To these devout petitions, Lord, give thou thy blessed grant! Amen."

The words, addressed by one potentate to another, with such a diplomatic mixture of flattering devotion and ornate self-confidence, were, apparently, exactly what were required. At any rate, the expedition was crowned with success. The secret of its purpose was well kept, and one day towards the end of June, 1596, the English armament suddenly appeared in the bay of Cadiz. At the first moment, an injudicious decision might have led to a disaster; the commanders had ordered a hazardous assault to be made by land; and it was only with difficulty that Raleigh persuaded them to change their plan and attack on the water. After that, all went swimmingly. "Entramos! Entramos!" shouted Essex, flinging his hat into the sea, as his ship sailed into the harbour. Within fourteen hours all was over. The Spanish fleet was destroyed and the town, with all its strength and riches, in the hands of the English. Among the Spaniards the disorganisation was complete; panic and folly had seized upon them. By a curious chance the Duke of Medina Sidonia was Governor of Andalusia. As if it were not enough to have led the Armada to its doom, it was now reserved for him to preside over the destruction of the most flourishing city of Spain. He hurried to the scene of action, wringing his hands in querulous despair. "This is shameful," he wrote to King Philip. "I told your Majesty how necessary it was to send me men and money, and I have never even received an answer. So now I am at my wit's end." He was indeed. The West Indian fleet of fifty merchantmen, laden with treasure worth eight million crowns, had fled into an inner harbour where it lay, in helpless confusion, awaiting its fate. Essex had ordered it to be seized, but there were delays among subordinates, and the unhappy Duke saw what must be done. He instantly gave commands; the whole fleet was set on fire; a faint smile, the first in seven years, was seen to flit across the face of Medina Sidonia; at last, in that intolerable mass of blazing ruin, he had got the better of his enemies.

While the honours of the sea-fight went to Raleigh, Essex was the hero on shore. He had led the assault on the city; his dash and bravery had carried all before them; and, when the victory was won, his humanity had put a speedy end to the excesses that were usual on such occasions. Priests and churches were spared; and three thousand nuns were transported to the mainland with the utmost politeness. The Spaniards themselves were in ecstasies over the chivalry of the heretic General. "Tan hidalgo," said Philip, "non ha vista entre herejas." The Lord Admiral himself was carried away with admiration. "I assure you," he wrote to Burghley, "there is not a braver man in the world than the Earl is; and I protest, in my poor judgment, a great soldier, for what he doth is in great order and discipline performed."

The English occupied Cadiz for a fortnight. Essex proposed that they should fortify the town an remain there until the Queen's pleasure was known When this was disallowed by the Council of War, he suggested a march into the interior of Spain; and, on this also being negatived, he urged that the flee should put out to sea, lie in wait for the returning

West Indian treasure-ships, and seize the vast booty they were bringing home. Once more he met with no support. It was decided to return to England immediately. A great ransom was raised from the inhabitants of Cadiz, the town was dismantled and destroyed, and the English sailed away. As they coasted back along the shores of Portugal, they could not resist a raid upon the unlucky town of Faro. The plunder was considerable, and it included one unexpected item—the priceless library of Bishop Jerome Osorius. The spectacle of so many marvellous volumes rejoiced the heart of the literary General; and he reserved them for himself, as his share of the loot. Yet, perhaps, he hardly glanced at them. Perhaps, as he sailed victoriously towards England, his wayward mind sank unexpectedly into an utterly incongruous mood. To be away from all this—and for ever! Away from the glory and the struggle—to be back at home, a boy again at Chartley—to escape irrevocably into the prolonged innocence of solitude and insignificance and dreams! With a play upon his own name—half smiling, half melancholy—he wrote some lines in which memory and premonition came together to give a strange pathos to the simple words.—

> Happy were he could finish forth his fate
> In some unhaunted desert, where, obscure
> From all society, from love and hate
> Of worldly folk, there should he sleep secure;

Then wake again, and yield God ever praise;
Content with hip, with haws, and brambleberry;
In contemplation passing still his days,
And change of holy thoughts to keep him merry:
Who, when he dies, his tomb might be the bush
Where harmless Robin resteth with the thrush:
—Happy were he!

VIII

On the same day on which Essex sailed from Cadiz, something of the highest moment was done in England: Elizabeth made Robert Cecil her Secretary, in name as well as in fact. That he had exercised the functions of the office for several years had not necessarily implied his continuance in that position. The Queen had been uncertain; the arrangement, she said, was temporary; there were other candidates for the post. Among these was Thomas Bodley, whose claims Essex had pushed forward with his customary vehemence—a vehemence which, once again, had failed in its effect. For Cecil was now definitely installed in that great office; all the outward prestige and all the inward influence that belonged to it were to be permanently his.

He sat at his table writing; and his presence was sweet and grave. There was an urbanity upon his features, some kind of explanatory gentleness, which, when he spoke, was given life and meaning by his exquisite elocution. He was all mild reasonableness—or so it appeared, until he left his chair, stood up, and unexpectedly revealed the stunted discomfort of deformity. Then another impression came upon one—

the uneasiness produced by an enigma: what could the combination of that beautifully explicit countenance with that shameful, crooked posture really betoken? He returned to the table, and once more took up his quill; all, once more, was perspicuous serenity. And duty too—that was everywhere—in the unhurried assiduity of the writing, the consummate orderliness of the papers and arrangements, the long still hours of expeditious toil. A great worker, a born administrator, a man of thought and pen, he sat there silent amid the loud violence about him—the *brio* of an Essex and a Raleigh, the rush and flutter of minor courtiers, and the loquacious paroxysms of Elizabeth. While he laboured, his inner spirit waited and watched. A discerning eye might have detected melancholy and resignation in that patient face. The spectacle of the world's ineptitude and brutality made him, not cynical—he was not aloof enough for that—but sad—was he not a creature of the world himself? He could do so little, so very little, to mend matters; with all his power and all his wisdom he could but labour, and watch, and wait. What else was possible? What else was feasible, what else was, in fact, anything but lunacy? He inspected the career of Essex with serious concern. Yet, perhaps, in some quite different manner, something, sometimes—very rarely—almost never—might be done. At a moment of crisis, a faint, a hardly perceptible impulsion might

ROBERT CECIL, EARL OF SALISBURY.

From the portrait at Woburn Abbey, by kind permission of the Duke of Bedford, K.G., K.B.E.

be given. It would be nothing but a touch, unbetrayed by the flutter of an eyelid, as one sat at table, not from one's hand, which would continue writing, but from one's foot. One might hardly be aware of its existence oneself, and yet was it not, after all, by such minute, invisible movements that the world was governed for its good, and great men came into their own?

That might be, in outline, the clue to the enigma; but the detailed working-out of the solution must remain, from its very nature, almost entirely unknown to us. We can only see what we are shown with such urbane lucidity—the devoted career of public service, crowned at last, so fortunately, by the final achievement—a great work accomplished, and the Earl of Salisbury supreme in England. So much is plain; but we are shown no more—no man ever was. The quiet minimum of action which led to such vast consequences is withdrawn from us. We can, with luck, catch a few glimpses now and then; but, in the main, we can only obscurely conjecture at what happened under the table.

Essex returned, triumphant and glorious. He was the hero of the hour. A shattering blow had been dealt to the hated enemy, and in the popular opinion it was to the young Earl, so daring, so chivalrous, so obviously romantic, that the victory was due. The old Lord Admiral had played no great part in the affair, and the fact that the whole expedition would have been a failure if the advice of Raleigh had not been followed at the critical moment was unknown. There seemed, in fact, to be only one person in England who viewed the return of the conqueror without enthusiasm; that person was the Queen. Never was the impossibility of foretelling what Elizabeth would do next more completely exemplified. Instead of welcoming her victorious favourite in rapturous delight, she received him with intense irritation. Something had happened to infuriate her; she had indeed been touched at a most sensitive point; it was a question of money. She had put down £50,000 for the expenses of the expedition, and what was she to get in return? Only, apparently, demands for more money, to pay the seamen's wages. It was, she declared, just as she had expected; she had foreseen it all; she had known from the very first that every one would make a fortune out of the business except herself. With infinite reluctance she disgorged another £2000 to keep the wretched seamen from starvation. But she would have it all back; and Essex should find that he was responsible. There certainly had been enormous leakages. The Spaniards themselves confessed to a loss of several millions, and the official estimate of the booty brought back to England was less than £13,000. Wild rumours were flying of the strings of pearls, the chains of gold, the golden rings and buttons, the chests of sugar, the casks of quicksilver, the damasks and the Portuguese wines, that had suddenly appeared in London. There were terrific wranglings at the council table. Several wealthy hostages had been brought back from Cadiz, and the Queen announced that all their ransoms should go into her pocket. When Essex protested that the soldiers would thereby lose their prize-money, she would not listen; it was only, she said, owing to their own incompetence that the loot had not been far greater; why had they not captured the returning West Indian fleet? The Cecils supported her with unpleasant questions. The new Secretary was particularly acid. Essex, who had good reason to expect a very different reception, was alternately depressed and exacerbated. "I see," he wrote to Anthony Bacon, "the fruits of these kinds of employments, and I assure you I am as much distasted with the glorious greatness of a favourite as I was before with the supposed happiness of a courtier, and call to mind the words of the wisest man that ever lived, who speaking of man's works crieth out, Vanity of vanities, all is but vanity." The Queen's displeasure was increased by another consideration. The blaze of popularity that surrounded the Earl was not to her liking. She did not approve of any one's being popular except herself. When it was proposed that thanksgiving services for the Cadiz victory should be held all over the country, her Majesty ordered that the celebrations should be limited to London. She was vexed to hear that a sermon had been preached in St. Paul's, in which Essex had been compared to the greatest generals of antiquity and his "justice, wisdom, valour and noble carriage" highly extolled; and she took care to make some biting remarks about his strategy at the next council. "I have a crabbed fortune that gives me no quiet," Essex wrote, "and the sour food I am fain still to digest may breed sour humours." It was an odd premonition; but he brushed such thoughts aside. In spite of everything he would struggle to keep his temper, and "as warily watch myself from corrupting myself as I do seek to guard myself from others."

His patience and forbearance were soon rewarded. News came that the West Indian fleet, laden with twenty million ducats, had entered the Tagus only two days after the English had departed. It seemed clear that if the plan urged by Essex had been adopted, that if the armament had waited off the coast of Portugal as he had advised, the whole huge treasure would have been captured. Elizabeth had a sudden revulsion. Was it possible that she had been unjust? Ungenerous? Certainly she had been misinformed. Essex swam up into high favour, and the Queen's anger, veering round full circle, was vented upon his enemies. Sir William Knollys, the Earl's uncle, was made a member of the Privy Council and

Comptroller of the Household. The Cecils were seriously alarmed, and Burghley, trimming his sails to the changing wind, thought it advisable, at the next council, to take the side of Essex in the matter of the Spanish ransoms. But the move was not successful. Elizabeth turned upon him in absolute fury. "My Lord Treasurer," she roared, "either for fear or favour, you regard my Lord of Essex more than myself. You are a miscreant! You are a coward!" The poor old man tottered away in a shaken condition to write a humble expostulation to the Earl. "My hand is weak, my mind troubled," he began. His case, he continued, was worse than to be between Scylla and Charybdis, "for my misfortune is to fall into both . . . . Her Majesty chargeth and condemneth me for favouring of you against her; your Lordship contrariwise misliketh me for pleasing of Her Majesty to offend you." He really thought that it was time for him to retire. "I see no possibility worthily to shun both these dangers, but by obtaining of licence to live an anchorite, or some such private life, whereunto I am meetest for my age, my infirmity, and daily decaying estate; but yet I shall not be stopped by the displeasure of either of you both to keep my way to heaven." Essex replied, as was fit, with a letter of dignified sympathy. But Anthony Bacon's comments were different; he did not conceal his delighted animosity. "Our Earl, God be thanked!"

he told a correspondent in Italy, "hath with the bright beams of his valour and virtue scattered the clouds and cleared the mists that malicious envy had stirred up against his matchless merit; which hath made the Old Fox to crouch and whine."

Burghley was indeed very much upset. He considered the whole situation carefully, and he came to the conclusion that perhaps, after all, he had made a mistake in his treatment of the Bacons. Would that young nobleman have ever reached so dangerous an eminence without the support of his nephews? Did not they supply him with just that intellectual stiffening, that background of sense and character, which his unstable temperament required? Was it possibly still not too late to detach them? He could but try. Anthony was obviously the more active and menacing of the two, and if he could be won over . . . He sent Lady Russell, the sister of his wife and Lady Bacon, on an embassy to her nephew, with conciliatory messages and bearing offers of employment and reward. The conversation was long, but it was fruitless. Anthony would not budge an inch. He was irrevocably committed to the Earl, whom he worshipped with the sombre passion of an invalid, his uncle's early neglect of him could never be forgiven or forgotten, and as for his cousin Robert, his hatred of him was only equalled by his scorn. He explained his feelings in detail to his aunt, who hardly knew what to answer. The Secretary, he declared, had actually "denounced a deadly feud" against him. "Ah, vile urchin!" said Lady Russell, "is it possible?" Anthony replied with a laugh and a Gascon proverb—"Brane d'âne ne monte pas al cicl." "By God," said Lady Russell, "but he is no ass." "Let him go for a mule then, Madam," rejoined Anthony, "the most mischievous beast that is." When his aunt had gone, Anthony wrote out a minute account of the conversation and sent it to his patron, concluding with a protestation to his "Good Lord" of "the entire devotion of my heart, together with the unchangeable vow of perfect obedience, which it hath long since no less resolutely than freely sworn unto your lordship, and the confidence I have in your lordship's most noble and true love." Why indeed should he change? How futile to suggest it! And now, when so many years of service had grown into adoration—now, when so many years of labour were blossoming into success!

For, in truth, the dreams of Anthony seemed to be on the brink of fulfilment; it was difficult to conceive what could prevent Essex from becoming before long the real ruler of England. His ascendancy over Elizabeth appeared to be complete. Her personal devotion had not lessened with time; on the contrary it seemed now to be reinforced by a growing recognition of his qualities as a soldier and a statesman. The

Cecils bowed before him; Raleigh was not admitted to the royal presence; no other rivals were visible. Dominating the council-table, he shouldered the duties and responsibilities of high office with vigour and assurance. Work poured in upon him; he had, he said, "to provide for the saving of Ireland, the contenting of France, the winning of the Low Countries to such conditions as they are yet far from; and the discovering and preventing of practices and designs, which are more and greater than ever." In the midst of so much business and so much success, he did not forget his friends. His conscience pricked him on the score of Thomas Bodley. What reparation could he make for the loss of the Secretaryship, which he had promised his faithful follower in vain? He bethought him of the library of Bishop Jerome Osorius, seized up so unexpectedly on that summer day at Faro. Bodley should have it—it was the very thing. And Bodley did have it; and such was the curious beginning of the great Library that bears his name.

Success, power, youth, royal favour, popular glory—what was lacking in the good fortune of the marvellous Earl? Only one thing, perhaps—and that too now was given him: the deathless consecration of Art. A supreme poet, blending together with the enchantment of words the loveliness of an hour and the vastness of human destiny, bestowed a splendid immortality upon the

"noble Peer,
Great England's glory and the world's wide wonder,
Whose dreadful name late through all Spain did
thunder,
And Hercules two pillars standing near
Did make to quake and fear.
Fair branch of Honour, flower of Chivalry,
That fillest England with thy triumph's fame,
Joy have thou of thy noble victory!"

The prowess and the person of Essex stand forth, lustrous and dazzling, before all eyes.

Yet there was one pair of eyes—and one only—that viewed the gorgeous spectacle without blinking. The cold viper-gaze of Francis Bacon, heedless of the magnificence of the exterior, pierced through to the inner quiddity of his patron's situation and saw there nothing but doubt and danger. With extraordinary courage and profound wisdom he chose this very moment—the apex, so it seemed, of Essex's career—to lift his voice in warning and exhortation. In a long letter, composed with elaborate solicitude and displaying at once an exquisite appreciation of circumstances, a consummate acquaintance with the conditions of practical life, and a prescience that was almost superhuman, he explained to the Earl the difficulties of his position, the perils that the future held in store for him, and the course of conduct by which those perils might be avoided. Everything, it was obvious, hinged upon the Queen; but Bacon perceived that in this very fact lay, not the strength, but the weakness of Essex's situation. He had no doubt what Elizabeth's half-conscious thoughts must be.—"A man of a nature not to be ruled; that hath the advantage of my affection, and knoweth it; of an estate not grounded to his greatness; of a popular reputation; of a militar dependence." What might not come of such considerations? "I demand," he wrote, "whether there can be a more dangerous image than this represented to any monarch living, much more to a lady, and of Her Majesty's apprehension?" It was essential that the whole of Essex's behaviour should be dominated by an effort to remove those suspicions from Elizabeth's mind. He was to take the utmost pains to show her that he was not "*opiniastre* and unrulable"; he was "to take all occasions, to the Queen, to speak against popularity and popular courses vehemently and to tax it in all others"; above all, he was utterly to eschew any appearance of "militar dependence." "Herein," wrote Bacon, "I cannot sufficiently wonder at your Lordship's course . . . for Her Majesty loveth peace. Next she loveth not charge. Thirdly, that kind of dependence maketh a suspected greatness." But there was more than that. Bacon clearly realised that Essex was not cut out to be a General; Cadiz, no doubt, had gone off well; but he distrusted these military excursions, and he urged the Earl to indulge in no more of them. There were rumours that he wished to be made the Master of the Ordnance; such thoughts were most unwise. Let him concentrate upon the Council; there he could control military matters without taking a hand in them; and, if he wished for a new office, let him choose one that was now vacant and was purely civilian in its character: let him ask the Queen to make him the Lord Privy Seal.

No advice could have been more brilliant or more pertinent. If Essex had followed it, how different would his history have been! But—such are the curious imperfections of the human intellect—while Bacon's understanding was absolute in some directions, in others it no less completely failed. With his wise and searching admonitions he mingled other counsel which was exactly calculated to defeat the end he had in view. Profound in everything but psychology, the actual steps which he urged Essex to take in order to preserve the Queen's favour were totally unfitted to the temperament of the Earl. Bacon wished his patron to behave with the Machiavellian calculation that was natural to his own mind. Essex was to enter into an elaborate course of flattery, dissimulation, and reserve. He was not in fact to imitate the subserviency of Leicester or Hatton—oh no!—but he was to take every opportunity of assuring Elizabeth that he followed these noblemen as patterns, "for I do not know a readier mean to make Her Majesty think you are in your right way." He must be very careful of his looks. If, after a dispute, he agreed that the Queen was right, "a man must not read formality in your countenance." And "fourthly, your Lordship should never be without some particulars afoot, which you should seem to pursue with earnestness and affection, and then let them fall, upon taking knowledge of Her Majesty's opposition and dislike." He might, for instance, "pretend a journey to see your living and estate towards Wales," and, at the Queen's request, relinquish it. Even the "lightest sort of particulars" were by no means to be neglected —"habits, apparel, wearings, gestures, and the like." As to "the impression of a popular reputation," that was "a good thing in itself," and besides "well governed, is one of the best flowers of your greatness both present and to come." It should be handled tenderly. "The only way is to quench it *verbis* and not *rebus*." The vehement speeches against popularity must be speeches and nothing more. In reality, the Earl was not to dream of giving up his position as the people's favourite. "Go on in your honourable commonwealth courses as before."

Such counsels were either futile or dangerous. How was it possible that the frank impetuosity of Essex should ever bend itself to these crooked ways? Every one knew—every one, apparently, but Bacon—

that the Earl was incapable of dissembling. "He can conceal nothing," said Henry Cuffe; "he carries his love and his hatred on his forehead." To such a temperament it was hard to say which was the most alien—the persistent practice of some profoundly calculated stratagem, or the momentary trickery of petty cunning. "Apparel, wearings, gestures!" How vain to hope that Essex would ever attend to that kind of tiresome particularity! Essex, who was always in a hurry or a dream—Essex, who would sit at table unconscious of what he ate or drank, shovelling down the food, or stopping suddenly to fall into some long abstraction—Essex, who to save his time would have himself dressed among a crowd of friends and suitors, giving, as Henry Wotton says, "his legs, arms, and breast to his ordinary servants to button and dress him, with little heed, his head and face to his barber, his eyes to his letters, and ears to petitioners," and so, clad in he knew not what, a cloak hastily thrown about him, would pass out, with his odd long steps, and his head pushed forward, to the Queen.

And, when he reached her, suppose that then, by some miracle, he remembered the advice of Bacon, and attempted to put into practice one or other of the contrivances that his friend had suggested. What would happen? Was it not clear that his nature would assert itself in spite of all his efforts?—that what was really in his mind would appear under his inexpert pretences, and his bungling become obvious to the far from blind Elizabeth? Then indeed his last state would be worse than his first; his very honesty would display his falsehood; and in his attempt to allay suspicions that were baseless he would actually have given them a reality.

Essex, no doubt, read and re-read Bacon's letter with admiration and gratitude—though perhaps, too, with some involuntary sighs. But he was soon to receive a very different admonition from another member of the family. Old Lady Bacon had been keeping, as was her wont, a sharp watch upon the Court from Gorhambury. Shortly after the Earl's return from Cadiz, she had received a surprisingly good report of his behaviour. He had suddenly—so Anthony wrote—given up his dissipated habits, and taken to "Christian zealous courses, not missing preaching or prayers in the Court, and showing true noble kindness towards his virtuous spouse, without any diversion." So far so good; but the amendment, it appeared, was not very lasting. Within a month or two, rumours were flying of an intrigue between the Earl and a married lady of high position. Lady Bacon was profoundly shocked; she was not, however, surprised; such doings were only to be expected in the godless world of London. The opportunity for a letter—a severely pious letter—presented itself. As for the lady in question, no words could be too harsh for such a creature. She was "unchaste and impudent, with, as it were, an incorrigible unshamefacedness." She was "an unchaste gaze and common by-word." "The Lord," she prayed, "speedily, by His grace, amend her, or"—that would be simplest—"cut her off before some sudden mischief." For Essex, such extreme measures were not yet necessary; he was, of course, less guilty, and there was still hope of his reformation. Let him read 1 Thess. iv. 3, and he would see that "this is the will of God, that ye should be holy, and abstain from fornication." Nay, more; he would find "a heavy threat that fornicators and adulterers God will judge, and that they shall be shut out; for such things, says the Apostle, commonly cometh the wrath of God upon us." Let him "take care, and 'grieve not the Holy Spirit of God.' " "With my very inward affection," she concluded, "have I thus presumed ill-favouredly to scribble, I confess, being sickly and weak in many ways."

Essex replied immediately, in the style of pathetic and dignified beauty that was familiar to him. "I take it," he wrote, "as a great argument of God's favour in sending so good an angel to admonish me; and of no small care in your Ladyship of my well-doing." He denied the whole story. "I protest before the majesty of God that this charge which is newly laid upon me is false and unjust; and that, since my departure from England towards Spain, I have been free from taxation of incontinency with any woman that lives." It was all, he declared, an invention of his enemies. "I live in a place where I am hourly conspired against, and practised upon. What they cannot make the world believe, that they persuade themselves unto; and what they cannot make probable to the Queen, that they give out to the world. . . . Worthy Lady, think me a weak man, full of imperfections; but be assured I do endeavour to be good, and had rather mend my faults than cover them." The Dowager did not quite know what to make of these protestations; perhaps they were genuine—she hoped so. He had begged her, in a postscript, to burn his letter; but she preferred not to. She folded it carefully up, with her crabbed fingers, and put it on one side, for future reference.

Whatever may have been the truth about the story that had reached her, it is clear that she no more understood the nature of her correspondent than she did that of her younger son. That devout austerity had too little in common with the generous looseness of the Earl, who, no doubt, felt that he might justly bow it on one side with some magnificent asseverations. His spirit, wayward, melancholy, and splendid, belonged to the Renaissance—the English Renaissance, in which the conflicting currents of ambition, learning, religion, and lasciviousness were so subtly intervolved. He lived and moved in a superb uncertainty. He did not know what he was or where he was going. He could not resist the mysterious dominations of moods—intense, absorbing, and utterly at variance with one another. He turned aside suddenly from the exciting whirl of business and politics to adore alone, in some inner room, the sensuous harmonies of Spenser. He dallied dangerously with court beauties; and then went to meditate for hours upon the attributes of the Deity in the cold church of St. Paul. His lot seemed to lead him irrevocably along the paths of action and power; and yet he could not determine whether that was indeed the true direction of his destiny; he dreamt of the remoteness of Lanfey and the serene solitudes of Chartley Chase. He was sent for by the Queen. He came into her presence, and another series of contradictory emotions overwhelmed him. Affection—admiration—exasperation—mockery—he felt them all by turns, and sometimes, so it seemed, simultaneously. It was difficult to escape the prestige of age, royalty, and success; it was impossible to escape the fascination of that rare intellect, with its alluring sinuosities and all the surprises of its gay vitality. His mind, swept along by hers, danced down delightful avenues. What happy twists! What new delicious vistas! And then—what had happened? The twists had grown abrupt, unaccountable, ridiculous. His head span. There was the way—plain and clear before them; but she insisted upon whisking round innumerable corners, and all his efforts could not keep her straight. She was a preposterous, obstinate old woman, fluctuating only when she should be firm, and strong in nothing but perversity. And he, after all, was a man, with a man's power of insight and determination; he could lead if she would follow; but Fate had reversed the rôles, and the natural master was a servant. Sometimes, perhaps, he could impose his will upon her—but after what an expenditure of energy, what a prolonged assertion of masculinity! A woman and a man! Yes, indeed, it was all too obvious! Why was he where he was? Why had he any influence whatever? It was not only obvious, it was ludicrous, it was disgusting: he satisfied the peculiar cravings of a virgin of sixty-three. How was this to end? His heart sank, and, as he was about to leave her, he caught sight of something inexplicable in those extraordinary eyes. He hurried home—to his wife, his friends, his sisters; and then, in his great house by the River, one of those physical collapses, which from his boyhood had never been long absent, would come upon him; incapable of thought or action, shivering in the agonies of ague, he would lie for days in melancholy and darkness upon his bed.

But, after all, he could not resist the pressure of circumstances, the nature of the time, the call to do and to lead. His vital forces returned to him, bringing with them the old excitements of adventure and jealousies of ambition. Spain loomed as ever upon the horizon; she had not been crushed at Cadiz; the snake was still dangerous, and must be scotched again. There was talk of another expedition. Francis Bacon might say what he would; but if there was one how would it be possible for the "noble Peer" of the Prothalamium to keep out of it? How could he leave the agitation and the triumph to Walter Raleigh? How could he stay behind with the hunchback secretary, writing at a table? In private, he pressed the Queen eagerly; and she seemed more amenable than usual; she agreed to the principle of an armed attack, but hesitated over its exact form. The news began to leak out, and Francis Bacon grew uneasy. The event, he saw, would show whether his advice was going to be taken: the parting of the ways was at hand.

In the meantime, while the future hung in the balance, that versatile intelligence was occupied in a different direction. In January, 1597, a small volume made its appearance—one of the most remarkable that has ever come from the press. Of its sixty pages, the first twenty-five were occupied by ten diminutive "Essays"—the word was new in English—in which the reflections of a matchless observer were expressed in an imperishable form. They were reflections upon the ways of this world, and particularly upon the ways of Courts. In later years Bacon enlarged the collection, widening the range of his subjects, and enriching his style with ornament and colour; but here all was terse, bare and practical. In a succession of gnomic sentences, from which every beauty but those of force and point had been strictly banished, he uttered his thoughts upon such themes as "Suitors," "Ceremonies and Respects," "Followers and Friends," "Expense," and "Negociating." "Some books," he wrote, "are to be tasted, others to be swallowed, and some few to be chewed and digested"; there can be no doubt to which category his own belongs. And, as one chews, one learns much, not only of the methods of politic behaviour, but of the nature of the author, and of that curious quality of mingled boldness and circumspection that was native to his mind. "Mean men must adhere," he says, in his essay on "Faction," "but great men that have strength in themselves were better to maintain themselves indifferent and neutral; yet," he adds, "even in beginners to adhere so moderately, as he be a man of the one faction, which is passablest with the other, commonly giveth best way." The book was dedicated to "Mr. Anthony Bacon, his dear brother"; but what did Anthony, with his instinct for uncompromising devotion, think of such an apophthegm?

Whatever Anthony might think, Francis could not help it; in the last resort he must be swayed not by his brother but by his perception of the facts. It was clear that one of those periodical crises, which seemed to punctuate the relations of the Queen and the Earl with ever-increasing violence, was rapidly approaching. It became known that a naval attack upon Spain had actually been decided upon; but who was to command it? Early in February, Essex took to his bed. The Queen came to visit him; he seemed to recover after so signal an act of favour; and then once more was prostrate. The nature of his ailment was dubious: was he sulking, or was he really ill? Perhaps he was both. For a fortnight he remained invisible, while the Queen fretted, and rumour after rumour flew round the Court. The signs of a struggle—a quarrel—were obvious. It was declared on good authority that the Queen had told him that he was to share the command of the expedition with Raleigh and Thomas Howard; and that thereupon the Earl had sworn to have nothing to do with it. At last Elizabeth's vexation burst out into speech. "I shall break him of his will," she exclaimed, "and pull down his great heart!" She wondered where he got his obstinacy; but, of course, it was from his mother—from Lettice Knowles, her cousin, that woman whom she hated—the widow of Leicester. Then the news came that the Earl was better, so much better that he had risen, and was about to depart from the Court immediately, to visit his estates in Wales.

Bacon could hardly doubt any more where all this was leading. He made up his mind. He was a beginner;

and it was for him "to adhere so moderately, as he be a man of the one faction, which is passablest with the other." He wrote to Burghley. He wrote with deliberation and subtle care. "I thought," he said, "it would better manifest what I desire to express, if I did write out of a settled consideration of mine own duty, rather than upon the spur of a particular occasion." He mingled flattery and gratitude, touching upon "your Lordship's excellent wisdom," and adding, "My singular good Lord, *ex abundantia cordis*, I must acknowledge how greatly and diversely your Lordship hath vouchsafed to tie me unto you by many your benefits." In a tone of deep respect and humility, he pressed his services upon his uncle. "This causeth me most humbly to pray your Lordship to believe that your Lordship is upon just title a principal owner and proprietary of that, I cannot call talent, but mite, that God hath given me; which I ever do and shall devote to your service." He even begged for forgiveness; he even dissociated himself—with an ameliorating parenthesis—from his brother Anthony. "In like humble manner I pray your Lordship to pardon mine errors, and not to impute unto me the errors of any other (which I know also themselves have by this time left and forethought); but to conceive of me to be a man that daily profiteth in duty." And he closed with a final protestation, cast in a sentence of superb rhythm, with a noble and touching fall. "And so again, craving your Honour's pardon for so long a letter, carrying so empty an offer of so unpuissant a service, but yet a true and unfeigned signification of an honest and vowed duty, I cease; commending your Lordship to the preservation of the Divine Majesty."

Burghley's answer is unknown to us; but we may be sure that he did not repel these advances, nor fail to note their implications. Events were now moving rapidly. The death of old Lord Cobham, by leaving vacant the Wardenship of the Cinque Ports, brought the crisis to a head. His son, the new Lord, hoped to succeed to the office; but he was hated by Essex, who pressed the claims of Sir Robert Sidney. For a week the conflict raged, and then the Queen announced her decision: the Wardenship should go to Lord Cobham. Thereupon Essex declared once more that he would leave the Court—that he had pressing business in Wales. All was prepared; men and horses were ready, and the Earl was only waiting to bid farewell to Burghley, when he was sent for by the Queen. There was a private interview, which ended in a complete reconciliation; and Essex emerged Master of the Ordnance.

So this was the consequence of Francis Bacon's advice! He had told the Earl to pretend a journey, in order to be able to waive it gracefully at the request of the Queen; and the foolish man had done the very opposite—had used it as a threat with which to force the royal hand. And to what end? To pursue what was most to be avoided—to emphasise that "militar dependence" which was at once so futile and so full of danger—nay, even to get possession of that very office, the Mastership of the Ordnance, which he had been particularly recommended to shun.

Clearly, the letter to Burghley was justified; it had become imperative for a "beginner" to acquire some other aid to the good things of this world besides what was offered by the dubious fortune of Essex. Yet it would be foolish to abandon the old connection altogether; it might still prove useful, in a variety of ways. For instance, Sir William Hatton was dead; he had left a rich widow—young and eligible; to marry her would be an excellent cure for that disease from which Bacon was still suffering—consumption of the purse. Negotiations were set on foot, and it seemed as if all might end happily, if the lady's father, Sir Thomas Cecil, could be brought to agree. Bacon begged Essex to use his influence; and Essex did all that he was asked. He wrote to Sir Thomas, expatiating upon the merits of his "dear and worthy friend" who, he had heard, was "a suitor to my Lady Hatton, your daughter." "To warrant my moving of you to incline favourably to his suit, I will only add this, that if she were my sister or daughter, I protest I would as confidently resolve myself to farther it, as now I persuade you. And though my love to him be exceedingly great, yet is my judgment nothing partial; for he that knows him so well as I do cannot but be so affected." Yet, once more, the Earl's influence was unavailing; for some unknown cause, Bacon was again disappointed; and Lady Hatton, like the Attorney-Generalship, went to Edward Coke.

Essex had not only been made Master of the Ordnance; he had also been given the command of the expedition against Spain. For months it had been known that the Spaniards had been busy with elaborate naval preparations in their great adjoining harbours of Corunna and Ferrol. The destination of the new Armada was unknown—perhaps it was Africa, or Brittany, or Ireland; but there were persistent reports that an attack was to be made on the Isle of Wight. It was decided to forestall the danger. Essex, with Raleigh and Lord Thomas Howard under him, was to take the fleet and a powerful armed force to Ferrol, and destroy all that he found there. The Cadiz adventure, in short, was to be repeated; and why not? The Queen herself believed that it might be done—cheaply, effectively and quickly. Even the Cecils agreed. Reconciliation was in the air. Burghley acted as peace-maker, and brought his son and the Earl together. Essex gave a little dinner at his house, to which was bidden not only Sir Robert, but Walter Raleigh as well. The enmities of years were laid aside; and, in a private conclave of two hours, the three great men bound themselves together in friendship. As a final proof of good-will, it was agreed that Elizabeth should be persuaded to take Raleigh once more into her favour. She yielded, readily enough, to the double pressure; he was summoned to her presence, graciously received, and told that he might resume his duties as Captain of the Guard. Raleigh celebrated the occasion by having made for him a suit of silver armour; and so once more, superb and glittering, the dangerous man stood in the royal antechamber at Whitehall.

And now it was summer, and the great fleet was almost ready to depart. Essex was on the coast, superintending the final preparations. He had taken his farewell of the Queen; but for a fortnight more he was in England, and the adieux were continued till the last moment in an impassioned correspondence. Difficulties, dangers, griefs there might be in that ambiguous relationship; but now absence seemed to make all things clear. Elizabeth was at her benignest. She sent off a stream of gifts and messages, she sent her portrait, she wrote constantly with her own hand. Essex was happy—active, important, excited; the great Queen, with all her majesty and all her affection, appeared before his imagination like some radiant fairy. She was his "most dear and most admired Sovereign." He could not express his feelings; but, since "words be not able to interpret for me, then to your royal dear heart I appeal, which, without my words, can fully and justly understand me. Heavens and earth shall witness for me. I will strive to be worthy of so high a grace and so blessed a happiness." He was tied to her "by more ties than ever was subject to a prince." His soul was "poured out with most earnest, faithful, and more than most affectionate wishes." He thanked her for her "sweet letters, indited by the spirit of spirits." She had heard a report that his ship leaked, and wrote to him in alarm to bid him take precautions against the danger. He was in Plymouth, on the eve of departure, when her letter reached him. "That infinite love," he wrote, "which I bear your Majesty makes me now love myself for your favour's sake; and therefore, be secure, dear Lady, that I will be as useful to bring myself home to you, as you would have me be." There was no danger, he assured her; the wind was favourable; all was ready; they were about to sail. "I humbly kiss your royal fair hands," he concluded, "and pour out my soul in passionate jealous wishes for all true joys to the dear heart of your Majesty, which must know me to be your Majesty's humblest and devoutest vassal, Essex." The fleet set out to sea.

IX

KING PHILIP sat working in the Escurial—the gigantic palace that he had built for himself, all of stone, far away, high up, amid the desolation of the rocky Guadarrama. He worked incessantly, as no monarch had ever worked before, controlling from his desk a vast empire—Spain and Portugal, half Italy, the Netherlands, the Western Indies. He had grown old and white-haired in his labours, but he worked on. Diseases had attacked him; he was tortured by the gout; his skin was cankered, he was the prey of a mysterious and terrible paralysis; but his hand moved over the paper from morning till night. He never emerged now. He had withdrawn into this inner room of his palace—a small room, hung with dark green tapestries—and there he reigned, secret, silent, indefatigable, dying. He had one distraction, and only one; sometimes he tottered through a low door into his oratory beyond and kneeling, looked out, through an inner window, as it were from a box of an opera, into the enormous spaces of a church. It was the centre of his great building, half palace and half monastery, and there, operatic too in their vestments and their movements and their strange singings, the priests performed at the altar close below him, intent upon their holy work. Holy! But his work too was that; he too was labouring for the glory of God. Was he not God's chosen instrument? The divine inheritance was in his blood. His father, Charles the Fifth, had been welcomed into Heaven, when he died, by the Trinity; there could be no mistake about it; Titian had painted the scene. He also would be received in a similar glorious fashion; but not just yet. He must finish his earthly duties first. He must make peace with France, he must marry his daughter, he must conquer the Dutch, he must establish everywhere the supremacy of the Catholic Church. There was indeed a great deal still to do, and very little time to do it in—he hurried back to his table; and it must all be done by himself, with his own hand.

His thoughts rushed round, confused and crowded. Not one was pleasant now. He had forgotten the fountains of Aranjuez and the eyes of the Princess of Eboli. Obscure incentives obsessed and agonised his brain—religion, pride, disappointment, the desire for rest, the desire for revenge. His sister of England rose before him—a distracting vision! He and she had grown old together, and she had always eluded him—eluded his love and his hate. But there was still just time; he would work more unrelentingly than ever before; and he would teach her—the unspeakable woman, with her heretic laughter—before he died, to laugh no longer.

That indeed would be a suitable offering with which to meet the Trinity. For years he had been labouring, with redoubled efforts, towards this end. His great Armada had not succeeded in its mission; that was true; but the reverse had not been an irreparable one. The destruction of Cadiz had also been unfortunate; but neither had that been fatal. Another Armada should be built and, with God's blessing, should achieve his purpose. Already he had accomplished much. Had he not been able, within a few months of the fall of Cadiz, to despatch a powerful fleet to Ireland, with a large army to succour the rebels there? It was unluckily a fact that the fleet had never reached Ireland, owing to a northerly gale, that more than twenty ships had sunk and that the remains of this second Armada had returned discomfited to Spain. But such accidents would happen, and why should he despair so long as the Trinity was on his side? With incredible industry he had set to work to have the fleet refitted in the harbour of Ferrol. He had put Martin de Padilla, the Governor (Adelantado) of Castile, in command of it, and Martin was a pious man, even more pious than Medina Sidonia. By the summer of 1597 it seemed as if the third Armada should be ready to start. Yet there were unaccountable delays. The Council sat in solemn conclave, but its elaborate discussions appeared, for some reason or other, not to help things forward. There were quarrels, too, among the commanders and officials; all were at loggerheads, without any understanding of the great task on which they were engaged. King Philip alone understood everything. His designs were his own secret; he would reveal them to no one; even the Adelantado, enquire as he would, should not be told the destination of the fleet. But there was to be no more of this procrastination. The Armada must sail at once.

Then came most disturbing news. The English fleet was being equipped; it was being assembled at Plymouth; very soon it would be on the high seas. And there could be little doubt of its objective; it would sail straight for Ferrol, and, once there—what was to prevent it?—the story of Cadiz would be repeated. The Adelantado declared that nothing could be done, that it was impossible to leave the harbour, that the preparations were altogether inadequate, that, in fact, he lacked *everything*, and could not face an enemy. It was exasperating—the pious Martin seemed to have caught Medina Sidonia's tone. But there was no help for it; one must face it out, and trust in the Trinity.

News came that the fleet had left Plymouth; and then—there was a miracle. After a terrifying pause it was known that a south-westerly gale had almost annihilated the English, whose ships, after ten days, had returned, with the utmost difficulty, into harbour. King Philip's Armada was saved.

The storm had indeed been an appalling one. The Queen in her palace had shuddered, as she listened to the awful wind; Essex himself had more than once given up his soul to God. His escape was less fortunate than he imagined; he was to be overwhelmed by a more terrible disaster; and the tempest was only an ominous prologue to the tragedy. With the fatal freshening of that breeze his good luck was over. From that moment misfortune steadily deepened upon him. By a curious coincidence the storm which ushered in such dreadful consequences has received a peculiar immortality. Among the young gentlemen who had sailed with the Earl in search of adventures and riches was John Donne. He suffered horribly, but he determined to convert his unpleasant sensations into something altogether unexpected. Out of the violence and disruption of a storm at sea he made a poem—a poem written in a new style and a new movement, without sensuous appeal or classic imagery, but harsh, modern, humorous, filled with surprising realistic metaphor and intricate wit.

"As sin-burdened souls from graves will creep
At the last day, some forth their cabins peep;
And tremblingly ask what news, and do hear so,
Like jealous husbands, what they would not know.

Some sitting on the hatches, would seem there
With hideous gazing to fear away fear.
Then note they the ship's sicknesses, the mast
Shaked with this ague, and the hold and waste
With a salt dropsy clogged, and all our tacklings
Snapping, like too high stretched treble strings;
And from our tattered sails, rags drop down so
As from one hanged in chains a year ago."

The verses, handed round everywhere in manuscript, were highly appreciated. It was the beginning of that extraordinary career of passion and poetry, which was to end in the fullness of time at the Deanery of St. Paul's.

While Donne was busy turning his acrobatic couplets, Essex was doing his utmost at Falmouth and Plymouth to repair the damage that had given rise to them. Commiseration came to him from Court. The Cecils wrote polite letters, and Elizabeth was in an unexpectedly gentle mood. "The Queen," Sir Robert told him, "is now so disposed to have us all love you, as she and I do talk every night like angels of you." An incident that had just occurred had so delighted her that she viewed the naval disaster with unusual equanimity. An Ambassador had arrived from Poland—a magnificent personage, in a long robe of black velvet with jewelled buttons, whom she received in state. Sitting on her throne, with her ladies, her counsellors, and her noblemen about her, she graciously gave ear to the envoy's elaborate harangue. He spoke in Latin; extremely well, it appeared; than, as she listened, amazement seized her. This was not at all what she had expected. Hardly a compliment—instead, protestations, remonstrances, criticisms—was it possible?—threats! She was lectured for presumption, rebuked for destroying the commerce of Poland, and actually informed that his Polish Majesty would put up with her proceedings no longer. Amazement gave way to fury. When the man at last stopped, she instantly leapt to her feet. "Expectavi orationem," she exclaimed, "mihi vero querelam adduxisti!"—and proceeded, without a pause, to pour out a rolling flood of vituperative Latin, in which reproof, indignation, and sarcastic pleasantries followed one another with astonishing volubility. Her eyes flashed, her voice grated and thundered. Those around her were in ecstasy; with all their knowledge of her accomplishments, this was something quite new—this prodigious power of *ex tempore* eloquence in a learned tongue. The unlucky ambassador was overwhelmed. At last, when she had rounded her last period, she paused for a moment, and then turned to her courtiers. "By God's death, my lords!" she said with a smile of satisfaction, "I have been enforced this day to scour up my old Latin which hath lain long rusting!" Afterwards she sent for Robert Cecil and told him that she wished Essex had been there to hear her Latin. Cecil tactfully promised that he would send the Earl a full account of what had passed; he did so, and the details of the curious scene have reached posterity, too, in his letter.

With some unwillingness she allowed the fleet to make another attack upon Spain. But it was now too weak to effect a landing at Ferrol; it must do no more than send fire-ships into the harbour in order to destroy the shipping; and after that an attempt might be made to intercept the West Indian treasure fleet. Essex set off with his diminished squadron, and once more the winds were against him. When, after great difficulty, he reached the Spanish coast, a gale from the East prevented his approaching the harbour of Ferrol. He wrote home, explaining his misadventure and announcing that, as he had received intelligence of the Spanish fleet having sailed to the Azores to meet the treasure transport, he intended to follow it thither immediately. Elizabeth sent him a reply, written in her most regal and enigmatic manner. "When I see," she said, "the admirable work of the Eastern wind, so long to last beyond the custom of nature, I see, as in a crystal, the right figure of my folly, that ventured supernatural haps upon the point of frenetical imputation." In other words, she realised that she was taking risks against her better judgment. She was like "the lunatic man that keeps a smack of the remains of his frenzy's freak, helped well thereto by the influence of *Sol in Leone*"—(it was August). Essex was not to presume too far on her unwise indulgence. She put in a "caveat, that this lunatic goodness make you not bold . . . to heap more errors to our mercy; . . . you vex me too much with small regard for what I scape or bid." He was to be cautious. "There remains that you, after your perilous first attempt, do not aggravate that danger with another in a farther-off climate, which must cost blows of good store; let character serve your turn, and be content when you are well, which hath not ever been your property." With a swift touch or two, delivered *de haut en bas*, she put her finger on his failings. "Of this no more, but of all my moods, I forget not my tenses, in which I see no leisure for ought but petitions, to fortify with best forwardness the wants of this army, and in the same include your safe return, and grant you wisdom to discern *verisimile* from *potest fieri*." And she concluded with an avowal of affection, in which the fullness of the feeling seems to be expressed by its very contortion. "Forget not to salute with my great favour good Thomas and faithful Mountjoy. I am too like the common faction, that forget to give thanks for what I received; but I was so loth to take that I had well nigh forgot to thank; but receive them now with millions and yet the rest keeps the dearest."

Her words went over the ocean to find him, and when they reached him it would have been well had he marked them more. At the Azores there was no sign of the Spanish fleet; but the treasure ships were expected to appear at any moment. Terceira, the central citadel in the Islands, was too strong to be attacked; and since, if the transport could once reach that harbour, it would be in safety, it was the plain policy of the English to lie in wait for it to the westward on the line of its route from America. It was decided to make a landing on the island of Fayal, which would be an excellent centre of observation. The whole fleet sailed towards it, but the ships failed to keep together, and when Raleigh's squadron reached the rendez-vous there was no sign of Essex or the rest. Raleigh waited for four days; then, being in want of water, he landed his men, attacked the town of Fayal, and took it. It was a successful beginning; Raleigh had commanded skilfully, and a good store of booty fell to him and his men. Immediately afterwards the rest of the fleet made its appearance. When Essex heard what had happened he was furious; Raleigh, he declared, had deliberately forestalled him for the sake of plunder and glory, and had disobeyed orders in attacking the island before the arrival of the commander-in-chief. The old quarrel flamed up sky-high. Some of Essex's more reckless partisans suggested to him that such an opportunity should not be missed—that Raleigh should be court-martialled and executed. Angry though Essex was, this was too much for him; "I would do it," he was reported to have said, "if he were my friend." At last an agreement was come to. It was arranged that Raleigh was to apologise, and that no mention of his successful action was to be recorded in the official report; he was to gain no credit for what he had done; on those conditions his misconduct would be passed over. There was a reconciliation, but Essex was still sore. So far he had done nothing worthy of his reputation—not a prize nor a prisoner was his. But he learnt that there was another island which might easily be captured; if Raleigh had taken Fayal, he would take San Miguel; and to San Miguel he instantly sailed. *Verisimile* and *potest fieri!* Why had he not marked those words? The attack upon San Miguel was an act of folly. For that island lay to the east of Terceira, and to go there was to leave the route of the treasure fleet unguarded. What might have been expected occurred. While the English were approaching San Miguel, the vast tribute of the Indies safely sailed into the harbour of Terceira. San Miguel after all proved to be so rocky as to make a landing impossible; Terceira was impregnable; all was over; there was nothing to be done but to return home.

Yes! But all this time where was the Spanish fleet? It had never left Ferrol, where the preparations of years were at last being completed with feverish rapidity. While King Philip was urging them forward in an endless stream of despatches, the news reached him that the English had sailed to the Azores. He saw that his opportunity had come. The odious island lay open and defenceless before him. Surely now his enemy was delivered into his hands. He ordered the Armada to sail immediately. It was in vain that the Adelantado begged for a little more delay, that he expatiated upon the scandalous deficiencies which made the armament unfit for service, that finally he implored to be relieved of his intolerable responsibility. In vain—the pious Martin, still ignorant of his destination, was forced to lead the fleet into the Bay of Biscay. Then, and only then, was he allowed to read his instructions. He was to sail straight for England, to attack Falmouth, to occupy it and, having defeated the enemy's fleet, to march towards London. The Armada sailed onwards, but as it approached Scilly a northerly wind fell upon it. The ships staggered and wavered; the hearts of the Captains sank. King Philip's preparations had been indeed inadequate; *everything*, as the Adelantado had said, was lacking—even elementary seamanship, even the desire to meet the foe. The spider of the Escurial had been spinning cobwebs out of dreams. The ships began to scatter and sink; the wind freshened to a gale; there was a despairing Council of War; the Adelantado gave the signal; and the Armada crept back into Ferrol.

King Philip was almost unconscious with anxiety and disease. He prayed incessantly, kneeling in anguish as he looked out from his opera box upon the high altar. Suddenly he was overwhelmed by a paralytic seizure; he hardly breathed, he could swallow no food, his daughter, hovering over him, blew liquid nourishment down his throat from a tube, and so saved his life. Already the news had come of the return of the Adelantado; but the King seemed to have passed beyond the reach of human messages. Suddenly there was a change; his eyes opened; he regained consciousness. "Will Martin never start?" were his first words. The courtiers had a painful task in front of them. They had to explain to King Philip that the pious Martin had not only started but that he had also come back.

X

ESSEX, too, had come back, and had to face a mistress who was by no means dying. A few Spanish merchantmen, accidentally picked up on the return journey, were all he could produce to justify an exploit which had not only been enormously expensive but had left England exposed to the danger of foreign invasion. Elizabeth had been unwilling to allow the fleet to depart after the great storm; she had been over-persuaded; and this was the consequence. Her rage was inevitable. Mismanagement—gross and inexcusable; severe loss, both of treasure and reputation; imminent peril to the realm: such was her summary of the business. The only compensation, she felt, was that she had now learnt her lesson. The whole policy, which she had always profoundly distrusted, of these dangerous and expensive expeditions, was finally shown to be senseless, and she would have no more of it. Never again, she declared to Burghley, would she send her fleet out of the Channel; and, for once in a way, she kept her word.

Received with icy disapprobation, Essex struggled to excuse himself, found that it was useless, and, mortified and angry, retired from the Court to the seclusion of his country house at Wanstead, on the eastern outskirts of London. From there he addressed a pathetic letter to the Queen. She had made him, he said, "a stranger," and "I had rather retire my sick body and troubled mind into some place of rest than, living in your presence, to come now to be one of those that look upon you afar off." "Of myself," he added, "it were folly to write that which you care not to know." Nevertheless, he assured her, "I do carry the same heart I was wont, though now overcome with unkindness, as before I was conquered by beauty. From my bed, where I think I shall be buried for some few days, this Sunday night. Your Majesty's servant, wounded but not altered by your unkindness, R. Essex."

"Conquered by beauty!" Elizabeth smiled, but she was not placated. What particularly annoyed her was to find that the popular reputation of the Earl as a great captain was in no way abated. The failure of the Islands Voyage was put down by the general public to ill luck, to the weather, to Raleigh—to every cause but the right one—the incompetence of the commander-in-chief. They were fools; and she knew where the truth lay. Yet she wished it were otherwise. One day, while she was expatiating on the theme in the garden at Whitehall, Sir Francis Vere ventured to speak up for the absent man. She listened graciously, argued a little, then changed her tone, and, leading

Sir Francis to the end of an alley, sat down with him and talked for a long time, with gentleness and affection, of Essex—his ways, his views, his curious character, his delightful disposition. Soon afterwards, she wrote to him, enquiring of his health. She wrote again, more pressingly. In her heart she wished him back, life was dull without him, the past might be forgotten. She wrote once more, with hints of forgiveness. "Most dear Lady," Essex replied, "your kind and often sending is able either to preserve a sick man, or rather to raise a man that were more than half dead to life again. Since I was first so happy as to know what love meant, I was never one day, nor one hour, free from hope and jealousy; and, as long as you do me right, they are the inseparable companions of my life. If your Majesty do in the sweetness of your own heart nourish the one, and in the justness of love free me from the tyranny of the other, you shall ever make me happy. . . . And so, wishing your Majesty to be mistress of that you wish most, I humbly kiss your fair hands."

She was charmed. Such protestations—all the more enticing for the very ambiguity of their phrasing—melted away the last remains of her resentment. He must come back immediately; and she prepared herself for a moving and thoroughly satisfactory scene of reconciliation.

But she was not to be happy so soon. When Essex

saw beyond a doubt that she wished him to return, he on his side grew remote and querulous. Surrounded by advisers less wise than Francis Bacon—his mother and his sisters, and the pushing military men who depended on his patronage—he allowed himself to listen to their suggestions and to begin playing a dubious game. The fact that he had failed indefensibly in the Islands Voyage only made him the more anxious to assert himself. His letters, written in a mixture of genuine regret and artful coquetry, had produced the desired effect. The Queen wished him back; very well, she might have her wish—but she must pay for it. He considered that on his part he had a serious grievance. Not only had Robert Cecil been made Chancellor of the Duchy of Lancaster in his absence, but, one week before his return, Lord Howard of Effingham had been created Earl of Nottingham. This was too much. The patent actually mentioned, among the reasons for this promotion, the capture of Cadiz; and all the world knew that the capture of Cadiz had been due to Essex alone. It was true that the patent also mentioned—naturally enough—the defeat of the Spanish Armada, that Howard was over sixty, and that an earldom seemed a fitting reward for his long and splendid career of public service. No matter, there was another more serious question at issue, and it was in fact as plain as day—so the hotheads assured themselves at Wanstead Park—that the whole affair had been arranged beforehand as a deliberate slight. Howard had already, before the Cadiz expedition, attempted, as Lord Admiral, to take precedence of Essex, who, as an Earl, had firmly resisted his pretensions. But now there could be no doubt about it: the Lord Admiral, if he was an Earl, took precedence by law of all other Earls—except the Great Chamberlain, the Lord Steward, and the Earl Marshal; and thus Essex would have to give place to this upstart Nottingham. Who could be surprised if, in these circumstances, he refused to return to Court? He declined to be insulted. If the Queen really wished to see him, let her make such an eventuality impossible; let her show the world, by some signal mark of her favour, that his position—so far from being weakened by the Islands Voyage—was more firmly established than ever.

It was announced that he was still far from well—that any movement from Wanstead was out of the question. Elizabeth loured. Her Accession Day—November 17th—was approaching, and the customary celebrations would lack something—decidedly they would lack something—in the absence of . . . but she refused to think of it. She grew restless, and a thunderstorm seemed to hang over the Court. The return of Essex was becoming of the highest importance to everybody. Lord Hunsdon addressed the Earl with a tactful remonstrance, but in vain. Then

Burghley wrote—not without humour. "By report," he said, "I hear that your Lordship is very sick, though, I trust, recoverable with warm diet." But Accession Day came and went without the presence of Essex. Burghley wrote again; even Nottingham sent a fine Elizabethan letter, protesting his friendship. He doubted "that some villanous device had been pursued to make your Lordship conceive ill of me: but, my Lord, if I have not dealt in all things concerning you, as I would have dealt withal had I been in your place, let me never enjoy the kingdom of Heaven!" Under this fusillade Essex weakened so far as to let it be known that he would return—if Her Majesty expressly required it. And then Elizabeth mounted her high horse. She would mention the matter no more; she had other things to think of; she must give the whole of her attention to the negotiations with the French Ambassador.

The French Ambassador did indeed require skilful handling. A new diplomatic situation was arising, so full of uncertainty that Elizabeth found it more difficult than ever to decide upon the course to take. King Philip had unexpectedly recovered after the return of his fleet to Ferrol. He had sent for the Adelantado, who, it was expected by the courtiers, would leave the King's presence for the gallows. But not at all; the interview was entirely devoted to a discussion of the forthcoming invasion of England, which was to take place in the Spring. There was to be a fourth Armada. Extraordinary efforts were to be made, the deficiencies of the past were to be rectified, and this time there would be no doubt of the result. A State paper was drawn up, to determine the steps which must be taken to ensure the success of the expedition. "The first," so ran this remarkable document, "is to recommend the undertaking to God, and to endeavour to amend our sins. But, since his Majesty has already given a general order to this effect, and has appointed a commander who usually insists upon this point, it will only be needful to take care that the order is obeyed and to promulgate it again." In the next place, a large sum of money must be raised, "with extraordinary rapidity and by every licit means that can be devised. In order to examine what means are licit, a committee of theologians must be summoned, to whom so great a matter may be confided, and their opinion should be adopted." Certainly, with such wisdom at the head of affairs, there could be no possible doubt whatever about the success of the scheme.

But, while the attack on England was maturing, King Philip was growing more and more anxious for peace with France. Henry IV was gradually establishing his position, and, when he recaptured Amiens, the moment for opening negotiations had come. The French King, on his side, wished for peace; he saw that he could obtain it; but, before coming to a conclusion, it was necessary to consult his two allies—the English and the Dutch. He hoped to persuade them to a general pacification, and with this end in view he despatched a special envoy, De Maisse, to London.

If De Maisse expected to extract a speedy reply to his proposals, he was doomed to disappointment. He was received at the English Court with respect and cordiality, but, as his questions grew more definite, the replies to them grew more vague. He had several interviews with Elizabeth, and the oracle was not, indeed, dumb; on the contrary, it was extremely talkative—upon every subject but the one in hand. The ambassador was perplexed, amazed, and fascinated, while the Queen rambled on from topic to topic, from music to religion, from dancing to Essex, from the state of Christendom to her own accomplishments. She touched upon King Philip, who, she said, had tried to have her murdered fifteen times. "How the man must love me!" she added with a laugh and a sigh. She regretted these fatal differences in religion, which, she considered, mostly turned upon bagatelles. She quoted Horace: "quidquid delirant reges, plectuntur Achivi." Yes, it was all too true; her people were suffering, and she loved her people, and her people loved her; she would rather die than diminish by one iota that mutual affection; and yet it could not last much longer, for she was on the brink of the grave. Then, before De Maisse could get in a word of expostulation, "No, no!" she exclaimed. "I don't think I shall die as soon as all that! I am not so old, *Monsieur l'Ambassadeur*, as you suppose."

The Queen's costumes were a source of perpetual astonishment to De Maisse, and he constantly took note of them in his journal. He learnt that she had never parted with a dress in the course of her life and that about three thousand hung in her wardrobes. On one occasion he experienced something more than astonishment. Summoned to an audience, he found Elizabeth standing near a window, in most unusual attire. Her black taffeta dress was cut in the Italian fashion, and ornamented with broad gold bands, the sleeves were open and lined with crimson. Below this dress which was open all down the front, was another of white damask, open also down to the waist; and below that again was a white chemise, also open. The amazed ambassador hardly knew where to look. Whenever he glanced at the Queen, he seemed to see far too much, and his embarrassment was still further increased by the deliberation with which, from time to time, throwing back her head as she talked, she took the folds of her dress in her hands and held them apart, so that, as he described it, "lui voyait-on tout l'estomac jusques au nombril." The costume was completed by a red wig, which fell on to her shoulders and was covered with magnificent pearls, while strings of pearls were twisted round her arms, and her wrists were covered with jewelled bracelets. Sitting down when he appeared, she discoursed for several hours with the utmost amiability. The Frenchman was convinced that she was trying to bewitch him; perhaps she was; or perhaps the unaccountable woman had merely been feeling a little vague and fantastic that morning when she put on her clothes.

The absence of Essex dominated the domestic situation, and De Maisse was not slow to perceive a state of tension in the atmosphere. The great Earl, hovering on the outskirts of London in self-imposed and ambiguous exile, filled every mind with fears, hopes, and calculations. The Queen's references to the subject, though apparently outspoken, were not illuminating. She assured the ambassador that if Essex had really failed in his duty during the Islands Voyage she would have cut off his head, but that she had gone into the question very thoroughly, and come to the conclusion that he was blameless. She appeared to be calm; her allusion to the Earl's execution seemed to be a piece of half-jocular bravado; and she immediately passed on to other matters. The courtiers were more agitated. There were strange rumours abroad. It was whispered that the Earl had announced his approaching departure for the West, and had declared that so many gentlemen were with him who had been ill-recompensed for their services that it would be dangerous to stay any longer near London. The rash remark was repeated everywhere by Essex's enemies; but it had no sequel, and he remained at Wanstead.

All through the month of December, while De Maisse was struggling to obtain some categorical pronouncement from Elizabeth, this muffled storm continued. At one moment Essex suggested that his difference with Nottingham might be settled by single combat—a proposal that, curiously enough, was not accepted. Nottingham himself grew testy, took to his bed, and talked of going into the country. At last, quite unexpectedly, Essex appeared at Court. It was instantly known that he had triumphed. On the 28th the Queen made him Earl Marshal of England. The office had been in abeyance for many years, and its revival and bestowal at this moment was indeed a remarkable sign of the royal favour; for the appointment automatically restored the precedency of Essex over Nottingham. Since the offices of Lord Admiral and Earl Marshal were by statute of equal rank, and since both were held by Earls, it followed that the first place belonged to him of the older creation.

A few days later De Maisse prepared to depart, having achieved nothing by his mission. He paid a visit of farewell to Essex, who received him with sombre courtesy. A great cloud, said the Earl, had been hanging over his head, though now it was melting away. He did not believe in the possibility of peace between Spain and England; but he was unwilling to take a part in those negotiations; it was useless—the Father and the Son alone were listened to. Then he paused, and added gloomily, "The Court is a prey to two evils—delay and inconstancy; and the cause is the sex of the sovereign." De Maisse, inwardly noting the curious combination of depression, anger, and ambition, respectfully withdrew.

The Earl might still be surly; but the highest of spirits possessed Elizabeth. The cruel suspense of the last two months—the longest and most anxious of those wretched separations—was over; Essex was back again; a new delightful zest came bursting into existence. France could wait. She would send Robert Cecil to talk to Henry. In the meantime—she looked gaily round for some object on which to vent her energy—yes, there was James of Scotland! That ridiculous young man had been up to his tricks again; but she would give him a lesson. It had come to her ears that he was actually sending out an envoy to the Courts of the Continent, to assert his right of succession to the English throne. His right of succession! It was positively a mania. He seemed to think she was already dead; but he would find he was mistaken. Lashing herself into a most exhilarating fury, she seized her pen, and wrote a letter to her brother of Scotland, well calculated to make him shake in his shoes. "When the first blast," she began, "of strange unused and seld heard-of sounds had pearsed my ears, I supposed that flyeing fame, who with swift quills ofte passeth with the worst, had brought report of some untrothe"; but it was not so. "I am sorry," she continued, "that you have so wilfully falen from your best stay, and will needs throwe yourself into the hurlpool of bottomless discredit. Was the haste soe great to hie to such oprobry? . . . I see well wee two be of very different natures . . . Shall imbassage be sent to forayne princes laden with instructions of your raishe advised charge? I assure you the travaile of your creased words shall passe the boundes of too many landes, with an imputation of such levytie, as when the true sonnshine of my sincere dealing and extraordinary care ever for your safety and honor shall overshade too far the dymme and mystie clowdes of false invectyves . . . And be assured, that you deale with such a kinge as will beare no wronges and indure no infamy. The examples have been so lately seen as they can hardly be forgotten, of a farr mightier and potenter prince than many Europe hath. Looke you not therefore without large amends I may or will slupper-up such indignities . . . And so I recomend you to a better mynde and more advysed conclusions."

Having polished off King James, she felt able to cope once more with King Henry. She told Robert Cecil that he should go to France as her special ambassador, and the Secretary was all assent and gratitude. Inwardly, however, he was uneasy; he did not relish the thought of a long absence abroad while the Earl remained at home in possession of the field; and, while he gravely sat over his dispatches, he wondered what could be done. He decided to be perfectly open—to approach his rival with a frank avowal of his anxieties. The plan worked; and Essex, in generous grandeur, remembering with a smile how, in his absence, both the Secretaryship and the Duchy of Lancaster had gone to Cecil, swore that he would steal no marches. Yet Cecil still felt uncomfortable. It happened that at that moment a large and valuable consignment of cochineal arrived from the Indies for the Queen. He suggested that Essex should be allowed the whole for £50,000, at the rate of eighteen shillings a pound, the market price of a pound of cochineal being between thirty and forty shillings; and he also recommended that Essex should be given £7000 worth of the precious substance as a free gift. Elizabeth readily consented, and the Earl found himself bound to the Secretary by something more than airy chivalry—by ties of gratitude for a very solid benefit.

Cecil had taken ship for France, when news of a

most alarming nature reached London. A Spanish fleet of thirty-eight fly-boats with 5000 soldiers on board was sailing up the Channel. Elizabeth's first thought was for her Secretary. She sent an urgent message, forbidding him to leave England; but he had already sailed, had missed the Spanish fleet, and arrived at Dieppe in safety. From there he at once despatched to his father a full account of the enemy's armament, writing on the cover of his letter, "For life, for life, for very life," with a drawing of a gallows, as a hint to the messenger of what would happen to him if he dallied on the road. There was not a moment's hesitation in London. The consultations of the Government were brief and to the point: orders were sent out in every direction, and no one asked the advice of the theologians. Lord Cumberland, with all the ships he could collect, was told to pursue the enemy; Lord Nottingham hurried to Gravesend, and Lord Cobham to Dover; Raleigh was commissioned to furnish provisions all along the coast; Essex was to stand ready to repel an attack wherever it might be delivered. But the alarm passed as quickly as it had arisen. Cumberland's squadron found the Spaniards outside Calais, and sank eighteen of the fly-boats; the rest of them huddled into the harbour, from which they never ventured to emerge.

Essex kept his promise. During the Secretary's absence, he supplied his place with the Queen, but made no attempt to take an unfair advantage of the situation. For the time indeed, his interests seemed to be elsewhere, and politics gave way to love-making. During the early wintry months of 1598 he kept himself warm at Court, philandering with the ladies. The rumours of his proceedings were many and scandalous. It was known that he had had a child by Mistress Elizabeth Southwell. He was suspected of a passion for Lady Mary Howard and of another for Mistress Russell. A court gossip reported it as certain that "his fairest Brydges" had once more captured the Earl's heart. While he passed the time with plays and banquets, both Lady Essex and the Queen were filled with uneasiness. Elizabeth's high spirits had suddenly collapsed; neither the state of Europe nor the state of Whitehall gave her any satisfaction; she grew moody, suspicious, and violent. For the slightest neglect, she railed against her Maids of Honour until they burst out crying. She believed that she had detected love-looks between Essex and Lady Mary Howard, and could hardly control her anger. She did, however, for the moment, privately determining to have her revenge before long. Her opportunity came when Lady Mary appeared one day in a particularly handsome velvet dress, with a rich border, powdered with pearl and gold. Her Majesty said nothing, but next morning she had the dress secretly abstracted from Lady Mary's wardrobe and brought to her.

That evening she electrified the Court by stalking in with Lady Mary's dress upon her; the effect was grotesque; she was far taller than Lady Mary and the dress was not nearly long enough. "Well, Ladies," she said, "how like you my new-fancied suit?" Then, amid the gasping silence, she bore down upon Lady Mary. "Ah, my Lady, and what think *you?* Is not this dress too short and ill-becoming?" The unfortunate girl stammered out an assent. "Why then," cried Her Majesty, "if it become not me, as being too short, am minded it shall never become thee, as being too ne; so it fitteth neither well"; and she marched out f the room again.

ch moments were disturbing; but Essex still had art to pacify the royal agitations. Then all was ance again, and spring was seen to be approaching and one could forget the perplexities of passion politics, and one could be careless and gay. In a cularly yielding moment, the Earl had persuaded Queen to grant him a great favour; she had agreed to see his mother—the odious Lettice Leicester, who had been banished from her presence for years. Yet, when it came to the point, Elizabeth jibbed. Time after time Lady Leicester was brought to the Privy Gallery; there she stood waiting for Her Majesty to pass; but, for some reason or other, Her Majesty always went out by another way. At last it was arranged that Lady Chandos should give a great dinner, at which the Queen and Lady Leicester should meet. Everything was ready; the royal coach was waiting; Lady Leicester stood at the entrance with a fair jewel in her hand, worth £300. But the Queen sent word that she should not go. Essex, who had been ill all day, got out of bed when he heard what had happened, put on a dressing-gown, and had himself conveyed to the Queen by a back way. It was all useless, the Queen would not move, and Lady Chandos's dinner party was indefinitely postponed. Then all at once Elizabeth relented. Lady Leicester was allowed to come to Court; she appeared before the Queen, kissed her hand, kissed her breast, embraced her, and was kissed in return. The reconcili tion was a very pretty one; but how long would the fair days last?

In the meantime, Cecil had failed as completely France as De Maisse in England. He returned, havi accomplished nothing, and early in May the ine table happened—Henry broke off from his allies, and by the treaty of Vervins, made peace with Spain. Elizabeth's comments were far from temperate. The French King, she said, was the Antichrist of Ingratitude; she had helped him to his crown, and now he had deserted her; it was true enough—but the wily Béarnais, like everybody else, was playing his own game. Burghley, however, was convinced that the situation required something more than vituperative outbursts. He wished for peace, and believed that it was still not too late to follow Henry's example; Philip, he thought, would be ready enough to agree to reasonable terms. Such were Burghley's views, and Essex violently opposed them. He urged an exactly contrary policy—a vigorous offensive—a great military effort, which would bring Spain to her knees. To start off with, he proposed an immediate attack upon the Indies; whereupon Burghley made a mild allusion to the Islands Voyage. And so began once more a long fierce struggle between the Earl and the Cecils—a struggle that turned the Council board into a field of battle, where the issues of Peace and War, the destinies of England, and the ambitions of hostile ministers jostled and hurtled together, while the Queen sat in her high chair at the head of the table, listening, approving, fiercely disagreeing, veering passionately from one side to the other, and never making up her mind.

Week after week the fight went on. Essex's strong card was Holland. Were we, he asked, to play the same trick on the Dutch as Henry had played on us? Were we to leave our Protestant allies to the tender mercy of the Spaniard? Burghley replied that the Dutch might join in a general pacification; and he countered Holland with Ireland. He pointed out that the only hope of effectually putting a stop to the running sore of Irish rebellion, which was draining the resources of England, was to make peace with Spain, whereby the rebels would be deprived of Spanish money and reinforcements, while at the same time England would be able to devote all her energies to a thorough conquest of the country. Current events gave weight to his words. The Lord Deputy Borough had suddenly died; there was confusion in Dublin; and Tyrone, the leader of the rebels in Ulster, had, after a patched-up truce, re-opened hostilities. In June it was known that he was laying siege to the fort on the river Blackwater, one of the principal English strongholds in the North of Ireland, and that the garrison was in difficulties. No new Lord Deputy had been appointed; who should be selected for that most difficult post? Elizabeth, gravely troubled, found it impossible to decide. It looked as if the Irish question was soon to become as intolerable as the Spanish one. As the summer days grew hotter, the discussions in the Council grew hotter too. There were angry explosions on either side. One day, after Essex had delivered a feverish harangue on his favourite topic—the infamy of a peace with Spain—Burghley drew out a prayer-book from his pocket and pointed with trembling finger to a passage in the fifty-fifth psalm. "Bloodthirsty and deceitful men," read Essex, "will not live out half their days." He furiously brushed aside the imputation; but everyone was deeply impressed; and there were some who recollected afterwards, with awe and wonder, the prophetic text of the old Lord Treasurer.

Essex felt that he was misunderstood, and composed a pamphlet to explain his views. It was a gallantly written work, but it convinced no one who was not convinced already. As for the Queen, she still wavered. The Dutch sent an embassy, offering large sums of money if she would continue the war. This was important, and she appeared to be coming round finally to an anti-Spanish policy; but it was appearance and nothing more; she sheered away again with utter indecision.

Nerves grew jangled, and tempers dangerously short. Everything, it was clear, was working up towards one of those alarming climaxes, with which all at Court had grown so familiar; and, while they waited in dread, sure enough, the climax came. But this time it was of a nature undreamt of by the imagination of any courtier: when the incredible story reached them, it was as if the earth had opened at their feet. The question of the Irish appointment had become pressing, and Elizabeth, feeling that something really must be done about it, kept reverting to the subject on every possible occasion, without any result. At last she thought she had decided that Sir William Knollys, Essex's uncle, was the man. She was in the Council Chamber, with Essex, the Lord Admiral, Robert Cecil, and Thomas Windebank, clerk of the signet, when she mentioned this. As often happened, they were all standing up. Essex, who did not want to lose the support of his uncle at Court, proposed instead Sir George Carew, a follower of the Cecils, whose absence in Ireland would, he thought, inconvenience the Secretary. The Queen would not hear of it, but Essex persisted; each was annoyed; they pressed their candidates; their words grew high and loud; and at last the Queen roundly declared that, say what he would, Knollys should go. Essex, overcome with irritation, contemptuous in look and gesture, turned his back upon her. She instantly boxed his ears. "Go to the devil!" she cried, flaring with anger. And then the impossible happened. The mad young man completely lost his temper, and, with a resounding oath, clapped his hand to his sword. "This is an outrage," he shouted in his sovereign's face, "that I will not put up with. I would not have borne it from your father's hands."—He was interrupted by Nottingham, who pressed him backwards. Elizabeth did not stir. There was an appalling silence; and he rushed from the room.

Unparalleled as was the conduct of Essex, there was yet another surprise in store for the Court, for the Queen's behaviour was no less extraordinary. She did nothing. The Tower—the block—heaven knows what exemplary punishment—might naturally have been expected. But nothing happened at all. Essex vanished into the country, and the Queen, wrapped in impenetrable mystery, proceeded with her usual routine of work and recreation. What was passing in her head? Had she been horrified into a paralysis? Was she overcome by the workings of outraged passion? Was she biding her time for some terrific revenge? It was impossible to guess. She swept on her way, until . . . there was indeed an interruption. The great, the inevitable, misfortune had come at last; Burghley was dying. Worn out by old age, the gout, and the cares of his great office, he was sinking rapidly to the grave. He had been her most trusted counsellor for more than forty years—from a time—how unbelievably distant!—when she had not been Queen of England. Her Spirit, she had always called him; and now her Spirit was leaving her for ever. She could attend to nothing else. She hoped against hope, she prayed, she visited him constantly, waiting with grand affection—the solicitude of some strange old fairy daughter—beside his dying bed. Sir Robert sent him game, but he was too feeble to lift the food to his mouth, and the Queen fed him herself. "I pray you," he wrote to his son, "diligently and effectually let Her Majesty understand how her singular kindness doth overcome my power to acquit it, who, though she will not be a mother, yet sheweth herself, by feeding me with her own princely hand, as a careful norice; and if I may be weaned to feed myself, I shall be more ready to serve her on the earth; if not, I hope to be, in heaven, a servitor for her and God's Church. And so I thank you for your partridges."

When all was over, Elizabeth wept long and bitterly; and her tears were still flowing—it was but ten days after Burghley's death—when yet another calamity fell upon her. There had been a terrible disaster in Ireland. Sir Henry Bagenal, marching at the head of a powerful army to the relief of the fort on the Blackwater, had been attacked by Tyrone; his army had been annihilated, and he himself killed. The whole of northern Ireland, as far as the walls of Dublin, lay open to the rebels. It was the most serious reverse that Elizabeth had suffered in the whole of her reign.

The news was quickly carried to Whitehall; it was also carried to the Escurial. King Philip's agony was coming to an end at last. The ravages of his dreadful diseases had overwhelmed him utterly; covered from head to foot with putrefying sores, he lay moribund in indescribable torment. His bed had been lifted into the oratory, so that his dying eyes might rest till the last moment on the high altar in the great church. He was surrounded by monks, priests, prayers, chantings, and holy relics. For fifty days and nights the extraordinary scene went on. He was dying as he had lived—in absolute piety. His conscience was clear: he had always done his duty; he had been infinitely industrious; he had existed solely for virtue and the glory of God. One thought alone troubled him: had he been remiss in the burning of heretics? He had burnt many, no doubt; but he might have burnt more. Was it because of this, perhaps, that he had not been quite as successful as he might have wished? It was certainly mysterious—he could not understand it—there seemed to be something wrong with his Empire—there was never enough money—the Dutch—the Queen of England . . . as he mused, a paper was brought in. It was the despatch from Ireland, announcing the victory of Tyrone. He sank back on his pillows, radiant; all was well, his prayers and his virtues had been rewarded, and the tide had turned at last. He dictated a letter to Tyrone of congratulation and encouragement. He promised immediate succour, he foretold the destruction of the heretics, and the ruin of the heretic Queen. A fifth Armada . . . he could dictate no more, and sank into a tortured stupor. When he awoke, it was night and there was singing at the altar below him; a sacred candle was lighted and put into his hand, the flame, as he clutched it closer and closer, casting lurid shadows upon his face; and so, in ecstasy and in torment, in absurdity and in greatness, happy, miserable, horrible, and holy, King Philip went off, to meet the Trinity.

XI

Essex had gone away to Wanstead, where he remained in a disturbed, uncertain, and unhappy condition. The alternating contradictions in his state of mind grew more extreme than ever. There were moments when he felt that he must fling himself at the feet of his mistress, that, come what might, he must regain her affection, her companionship, and all the sweets of the position that had so long been his. He could not—he would not—think that he had been in the wrong; she had treated him with an indignity that was unbearable; and then as he brooded over what had happened, anger flamed up in his heart. He would tell her what he thought of her. Had he not always done so—ever since that evening, more than ten years ago, when he had chided her so passionately, with Raleigh standing at the door? He would chide her now, no less passionately, but, as was fitting, in a deeper and a sadder tone. "Madam," he wrote, "when I think how I have preferred your beauty above all things, and received no pleasure in life but by the increase of your favour towards me, I wonder at myself what cause there could be to make me absent myself one day from you. But when I remember that your Majesty hath, by the intolerable wrong you have done both me and yourself, not only broken all laws of affection, but done against the honour of your sex, I think all places better than that where I am, and all dangers well undertaken, so I might retire myself from the memory of my false, inconstant and beguiling pleasures. . . . I was never proud, till your Majesty sought to make me too base. And now, since my destiny is no better, my despair shall be as my love was, without repentance. . . . I must commend my faith to be judged by Him who judgeth all hearts, since on earth I find no right. Wishing your Majesty all comforts and joys in the world, and no greater punishment for your wrongs to me, than to know the faith of him you have lost, and the baseness of those you shall keep,

"Your Majesty's most humble servant,

"R. ESSEX."

When the news of the disaster on the Blackwater reached him, he sent another letter, offering his services, and hurried to Whitehall. He was not admitted. "He hath played long enough upon me," Elizabeth was heard to remark, "and now I mean to play awhile upon him, and stand as much upon my greatness as he hath upon stomach." He wrote a long letter of expostulation, with quotations from Horace, and vows of duty. "I stay in this place for no other purpose but to attend your commandment." She sent him a verbal message in reply. "Tell the Earl that I value *my*self at as great a price as *he* values *him*self." He wrote again: "I do confess that, as a man, I have been more subject to your natural beauty than as a subject to the power of a king." He obtained an interview; the Queen was not ungracious; the onlookers supposed that all was well again. But it was not, and he returned to Wanstead in darker dudgeon than ever.

It was clear that what Elizabeth was waiting for was some apology. Since this was not forthcoming, a deadlock had apparently been reached, and it seemed to the moderate men at Court that it was time an effort should be made to induce the Earl to realise the essence of the situation. The Lord Keeper Egerton, therefore, composed an elaborate appeal. Did not Essex understand, he asked, that his present course was full of danger? Did he not see that he was encouraging his enemies? Had he forgotten his friends? Had he forgotten his country? There was only one thing to do—he must beg for the Queen's forgiveness; whether he was right or wrong could make no difference. "Have you given cause, and yet take scandal to yourself? Why then, all you can do is too little to give satisfaction. Is cause of scandal given to you? Let policy, duty, and religion enforce you to yield, and submit to your sovereign, between whom and you there can be no proportion of duty." "The difficulty, my good Lord," Egerton concluded, "is to conquer yourself, which is the height of all true valour and fortitude, whereunto all your honourable actions have tended. Do it in this, and God will be pleased, Her Majesty well satisfied, your country will take good, and your friends comfort by it; yourself shall receive honour; and your enemies, if you have any, shall be disappointed of their bitter-sweet hope."

Essex's reply was most remarkable. In a style no less elaborate than the Lord Keeper's, he rebutted all his arguments. He denied that he was doing wrong either to himself or his friends; the Queen's conduct, he said, made it impossible for him to act in any other way. How could he serve his country when she had "driven him into a private kind of life"—when she had "dismissed, discharged, and disabled" him? "The indissoluble duty," he continued, "which I owe to Her Majesty is only the duty of allegiance, which I never will, nor never can, fail in. The duty of attendance is no indissoluble duty. I owe to Her Majesty the duty of an Earl and Lord Marshal of England. I have been content to do Her Majesty the service of a clerk, but can never serve her as a villain or a slave." As he wrote, he grew warmer. "But, say you, I must yield and submit; I can neither yield myself to be guilty, or this imputation laid upon me to be just. . . . Have I given cause, ask you, and take scandal when

I have done? No, I give no cause. . . . I patiently bear all, and sensibly feel all, that I then received when this scandal was given me. Nay more"—and now he could hold himself in no longer—"when the vilest of all indignities are done unto me, doth religion enforce me to sue?" The whole heat of his indignation was flaring out. "Doth God require it? Is it impiety not to do it? What, cannot princes err? Cannot subjects receive wrong? Is an earthly power or authority infinite? Pardon me, pardon me, my good Lord, I can never subscribe to these principles. Let Solomon's fool laugh when he is stricken; let those that mean to make their profit of princes shew to have no sense of princes' injuries; let them acknowledge an infinite absoluteness on earth, that do not believe in an absolute infiniteness in heaven. As for me, I have received wrong, and feel it. My cause is good, I know it; and whatsoever come, all the powers on earth can never shew more strength and constancy in oppressing than I can shew in suffering whatsoever can or shall be imposed on me."

Magnificent words, certainly, but dangerous, portentous, and not wise. What good could come of flaunting republican sentiments under the calm nose of a Tudor? Such oratory was too early or too late. Hampden would have echoed it; but in truth it was the past rather than the future that was speaking with the angry pen of Robert Devereux. The blood of a hundred Barons who had paid small heed to the

Lord's Anointed was pulsing in his heart. Yes! If it was a question of birth, why should the heir of the ancient aristocracy of England bow down before the descendant of some Bishop's butler in Wales? Such were his wild feelings—the last extravagance of the Middle Ages flickering through the high Renaissance nobleman. The facts vanished; his outraged imagination preferred to do away with them. For, after all, what had actually happened? Simply this, he had been rude to an old lady, who was also a Queen, and had had his ears boxed. There were no principles involved, and there was no oppression. It was merely a matter of bad temper and personal pique.

A realistic observer would have seen that in truth there were only two alternatives for one in Essex's position—a graceful apology followed by a genuine reconciliation with the Queen, or else a complete and final retirement from public life. More than once his mind swayed—as so often before—towards the latter solution; but he was not a realist, he was a romantic—passionate, restless, confused, and he shut his eyes to what was obvious—that, as things stood, if he could not bring himself to be one of those who "make their profit of princes" he must indeed make up his mind to a life of books and hunting at Chartley. Nor were those who surrounded him any more realistic than himself. Francis Bacon had for many months past avoided his company; Anthony was an enthusiastic devotee; Henry Cuffe was rash and cynical; his sisters were too ambitious, his mother was too much biased by her lifelong quarrel with Elizabeth, to act as a restraining force. Two other followers completed his intimate domestic circle. His mother's husband—for Lady Leicester had married a third time—was Sir Christopher Blount. A sturdy soldier and a Roman Catholic, he had served his stepson faithfully for many years, and, it was clear, would continue to do so, whatever happened, to the end. More dubious, from every point of view, was the position of Charles Blount, Lord Mountjoy. The tall young man with the brown hair and the beautiful complexion, who had won Elizabeth's favour by his feats at tilting, and who had fought a duel with Essex over the golden chessman given him by the Queen, had grown and prospered with the years. The death of his elder brother had brought him the family peerage; he had distinguished himself as Essex's lieutenant in all his expeditions, and he had never lost the favour of Elizabeth. But he was united to Essex by something more than a common military service—by a singular romance. The Earl's favourite sister, Lady Penelope, had been the Stella Sir Philip Sidney had vainly loved. She had married Lord Rich, while Sidney had married Walsingham's daughter, who, on Sir Philip's death, had become the wife of Essex. Penelope had not been happy; Lord Rich was an odious husband,

and she had fallen in love with Lord Mountjoy. A liaison sprang up—a lifelong liaison—one of those indisputable and yet ambiguous connexions which are at once recognised and ignored by society—between Essex's friend and Essex's sister. Thus Mountjoy, doubly bound to the Earl, had become—or so it seemed—the most faithful of his adherents. The little group—Essex, Lady Essex, Mountjoy, and Penelope Rich—was held together by the deepest feelings of desire and affection; while behind and above them all there hovered, in sainted knightliness, the shade of Sir Philip Sidney.

And so there was no barrier to hold Essex back from folly and intemperance; on the contrary, the characteristics of his environment—personal devotion, family pride, and military zeal—all conspired to urge him on. More remote influences worked in the same direction. Throughout the country the Earl's popularity was a growing force. The reasons for this were vague, but none the less effectual. His gallant figure had taken hold of the popular imagination; he was generous and courteous; he was the enemy of Raleigh, who was everywhere disliked; and now he was out of favour and seemed to be hardly used. The puritanical City of London, especially, tending, as it always did, to be hostile to the Court, paid an incongruous devotion to the unregenerate Earl. The word went round that he was a pillar of Protestantism, and

Essex, who was ready enough to be all things to all men, was not unwilling to accept the *rôle*. Evidence of another kind of esteem appeared when, on the death of Burghley, the University of Cambridge at once elected him to fill the vacant place of Chancellor. He was delighted by the compliment, and as a mark of gratitude presented the University with a silver cup of rare design. The curious goblet still stands on the table of the Vice-Chancellor, to remind the passing generations of Englishmen at once of the tumult of the past and of the placid continuity of their history.

Egged on by private passion and public favour, the headstrong man gave vent, in moments of elation, to strange expressions of anger and revolt. Sir Christopher Blount was present at Wanstead when one of these explosions occurred, and, though his stepson's words were whirling and indefinite, they revealed to him with startling vividness a state of mind that was full, as he said afterwards, of "dangerous discontentment." But the moments of elation passed, to be succeeded by gloom and hesitation. What was to be done? There was no satisfaction anywhere; retirement, submission, defiance—each was more wretched than the others; and the Queen still made no sign.

In reality, of course, Elizabeth too was wavering. She kept up a bold front; she assured everybody, including herself, that this time she was really going to be firm; but she knew well enough how many times before she had yielded in like circumstances, and experience indicated that the future would resemble the past. As usual, the withdrawal of that radiant presence was becoming insupportable. She thought of Wanstead—so near, so far—and almost capitulated. Yet no, she would do nothing, she would go on waiting; only a little longer, perhaps, and the capitulation would come from the other side. And then one dimly discerns that, while she paused and struggled, a new and a sinister element of uncertainty was beginning to join the others to increase the fluctuation of her mind. At all times she kept her eyes and her ears open; her sense of the drifts of feeling and opinion was extremely shrewd, and there were many about her who were ready enough to tell unpleasant stories of the absent favourite and expatiate on his growing—his extraordinary—popularity all over the country. One day a copy of the letter to Egerton was put into her hand. She read it, and her heart sank; she scrupulously concealed her feelings, but she could no longer hide from herself that the preoccupation which had now come to wind itself among the rest that perturbed her spirit was one of alarm. If that was his state of mind—if that was his position in the country . . . she did not like it at all. The lion-hearted heroine of tradition would not have hesitated in such circumstances—would have cleared up the situation in one bold and final stroke. But that was very far indeed from being Elizabeth's way. "Pusillanimity," the Spanish ambassadors had reported; a crude diagnosis; what really actuated her in the face of peril or hostility was an innate predisposition to hedge. If there was indeed danger in the direction of Wanstead she would not go out to meet it—oh no!—she would propitiate it, she would lull it into unconsciousness, she would put it off, and put it off. That was her instinct; and yet, in the contradictory convolutions of her character, another and a completely opposite propensity may be perceived, which nevertheless—such is the strange mechanism of the human soul—helped to produce the same result. Deep in the recesses of her being, a terrific courage possessed her. She balanced and balanced, and if, one day, she was to find that she was exercising her prodigies of agility on a tight-rope over an abyss—so much the better! She knew that she was equal to any situation. All would be well. She relished everything—the diminution of risks and the domination of them; and she would proceed, in her extraordinary way, with her life's work, which consisted . . . of what? Putting out flames? Or playing with fire? She laughed; it was not for her to determine!

Thus it happened that when the inevitable reconciliation came it was not a complete one. The details are hidden from us; we do not know the terms of the peace; we only know that the pretext for it was yet another misfortune in Ireland. Sir Richard Bingham had been sent out to take command of the military operations, and early in October, immediately upon his arrival at Dublin, he died. All was in confusion once more; Essex again offered his services; and this time they were accepted. Soon the Queen and the favourite were as much together as they had ever been. It appeared that the past had been obliterated, and that the Earl—as was his wont—had triumphantly regained his old position, as if there had never been a quarrel. In reality it was not so; the situation was a new one; mutual confidence had departed. For the first time, each side was holding something back. Essex, whatever his words, his looks, and even his passing moods may have been, had not uprooted from his mind the feelings of injury and defiance that had dictated his letter to Egerton. He had returned to Court as unchastened and undecided as ever, blindly impelled by the enticement of power. And Elizabeth on her side had by no means forgotten what had happened; the scene in the Council Chamber still rankled; she perceived that there was something wrong with those protestations; and, while she conversed and flirted as of old, she kept open a weather eye.

But these were subtleties it was very difficult to make sure of, as the days whirled along at Whitehall and Greenwich and Nonesuch; and even Francis Bacon could not quite decide what had occurred.

Possibly Essex was really again in the ascendant; possibly, after the death of Burghley, the star of Cecil was declining; it was most unwise to be too sure. For more than a year, gradually moving towards the Cecils, he had kept out of the Earl's way. In repeated letters he had paid his court to the Secretary, and his efforts had at last been rewarded in a highly gratifying manner. A new assassination plot had come to light—a new Catholic conspiracy; the suspects had been seized; and Bacon was instructed to assist the Government in the unravelling of the mystery. The work suited him very well, for, while it provided an excellent opportunity for the display of intelligence, it also brought him into a closer contact with great persons than he had hitherto enjoyed. And it turned out that he was particularly in need of such support. He had been unable to set his finances in order. The Mastership of the Rolls and Lady Hatton had both eluded him; and he had been obliged to content himself with the reversion to the Clerkship of the Star Chamber—with the prospect, instead of the reality, of emolument. Yet it had seemed for a moment as if the prospect were unexpectedly close at hand. The actual Clerk was accused of peculation, and the Lord Keeper Egerton was appointed, with others, to examine into the case. If the Clerk were removed, Bacon would succeed to the office; he wrote a secret letter to Egerton; he promised, in that eventuality, to resign the office to Egerton's son, on the understanding that the Lord Keeper on his side would do his best to obtain for him some compensating position. The project failed, for the Clerk was not removed, and Bacon did not come into his reversion for ten years. In the meantime, an alarming poverty stared him in the face. He continued to borrow—from his brother, from his mother, from Mr. Trott; the situation grew more and more serious; at last, one day, as he was returning from the Tower after an examination of the prisoners concerned in the assassination plot, he was positively arrested for debt. Robert Cecil and Egerton, however, to whom he immediately applied for assistance, were able between them to get him out of this difficulty, and his public duties were not interrupted again.

But, if the Secretary was useful, the Earl might be useful too. Now that he was back at Court, it would be well to write to him. "That your Lordship," Bacon said, "is *in statu quo primo* no man taketh greater gladness than I do; the rather because I assure myself that of your eclipses, as this hath been the longest, it shall be the last." He hoped that "upon this, experience may found more perfect knowledge, and upon knowledge more true consent. . . . And therefore, as bearing unto your Lordship, after her Majesty, of all public persons the second duty, I could not but signify unto you my affectionate gratulation."[2]

So far so good; but now the clouds of a new tempest were seen to be gathering on the horizon, filling the hearts of the watchers at Whitehall with perplexity and perturbation. It was absolutely necessary that some one should be made Lord Deputy of Ireland. After the shattering scene in the summer, nothing had been done; the question was urgent; upon its solution so much, so very much, depended! The Queen believed that she had found the right man—Lord Mountjoy. Besides admiring his looks intensely, she had a high opinion of his competence. He was approached on the subject, and it was found that he was willing to go. For a short time it appeared that the matter was happily settled—that Mountjoy was the *deus ex machina* who would bring peace not only to Ireland but to Whitehall. But again the wind shifted. Essex once more protested against the appointment of one of his own supporters; Mountjoy, he declared, was unfit for the post—he was a scholar rather than a general. It looked as if the fatal round of refusal and recrimination was about to begin all over again. Who then, Essex was asked, did he propose? Some year before, Bacon had written him a letter of advice precisely on this affair of Ireland. "I think," said the man of policy, "if your Lordship lent your reputation in this case—that is, to *pretend* that you would accept the charge—I think it would help you to settle Tyrone in his seeking accord, and win you a great deal of honour *gratis*." There was only one objection, Bacon thought, to this line of conduct: "Your Lordship is too quick to pass in such cases from dissimulation to verity." We cannot trace all the moves—complicated, concealed, and fevered—that passed at the Council table; but it seems probable that Essex, when pressed to name a substitute for Mountjoy, remembered Bacon's advice. He gave it as his opinion, Camden tells us, that "into Ireland must be sent some prime man of the nobility which was strong in power, honour, and wealth, in favour with military men, and which had before been general of an army; so as he seemed with the finger to point to himself." The Secretary, with his face of gentle conscientiousness, sat silent at the Board. What were his thoughts? If the Earl were indeed to go to Ireland—it would be a hazardous decision; but if he himself wished it—perhaps it would be better so. He scrutinised the future, weighing the possibilities with deliberate care. It was conceivable that the Earl, after all, was dissembling, that he understood how dangerous it would be for him to leave England, and was only making a show. But Cecil knew, as well as his cousin, the weak places in that brave character—knew the magnetism of arms and action—knew the tendency "to pass from dissimulation to verity." He thought he saw what would happen. "My Lord Mountjoy," he told a confidential correspondent, "is named; but to you, in secret I speak of it, not as a secretary but as a friend, that I think the Earl of Essex shall go Lieutenant of the Kingdom." He sat writing; we do not know of his other faint imperceptible movements. We only know that, in the Council, there were some who still pressed for the appointment of Mountjoy, that the Earl's indication of himself was opposed or neglected, and that then the candidature of Sir William Knollys was suddenly revived.

Opposition always tended to make Essex lose his head. He grew angry; the Mountjoy proposal seriously vexed him, and the renewal of Knollys' name was the last straw. He fulminated against such notions, and, as he did so, slipped—after what he had himself said, it was an easy, an almost inevitable transition—into an assertion of his own claims. Some councillors supported him, declaring that all would be well if the Earl went; the Queen was impressed; Essex had embarked on a heated struggle—he had pitted himself against Knollys and Mountjoy, and he would win. Francis Bacon had prophesied all too truly—the reckless man had indeed "passed from dissimulation to verity." Win he did. The Queen, bringing the discussion to a close, announced her decision: since Essex was convinced that he could pacify Ireland, and since he was so anxious for the office, he should have it; she would make him her Lord Deputy. With long elated strides and flashing glances he left the room in triumph; and so—with shuffling gait and looks of mild urbanity—did Robert Cecil.

It was long before Essex began to realise fully what had happened. The sense of victory, both at the moment and in anticipation—both at home and in Ireland—buoyed him up and carried him forward. "I have beaten Knollys and Mountjoy in the Council," he wrote to his friend and follower, John Harington, "and by God I will beat Tyrone in the field; for nothing worthy Her Majesty's honour hath yet been achieved."

Naturally enough the old story was repeated, and the long, accustomed train of difficulties, disappointments, and delays dragged itself out. Elizabeth chaffered over every detail, changed from day to day the size and nature of the armament that she was fitting out, and disputed fiercely upon the scope of the authority with which the new Lord Deputy was to be invested. As the weeks passed in angry bickering Essex sank slowly downwards from elation to gloom. Perhaps he had acted unwisely; regrets attacked him; the future was dark and difficult; what was he heading for? He was overwhelmed by miserable sensations; but it was too late now to draw back, and he must face the inevitable with courage. "Into Ireland I go," he told the young Earl of Southampton, who had become his devoted disciple; "the Queen hath irrevocably decreed it, the Council do passionately urge it, and I am tied in my own reputation to use no tergiversation; and, as it were indecorum to slip collar now, so would it also be *minime tutum;* for Ireland would be lost, and though it perished by destiny I should only be accused of it, because I saw the fire burn and was called to quench it, but gave no help." He was well aware, he said, of the disadvantages of absence—"the opportunities of practising enemies" and "the construction of Princes, under whom *magna fama* is more dangerous than *mala*." He realised and enumerated the difficulties of an Irish campaign. "All these things," he declared, "which I am like to see, I do now foresee." Yet to every objection he did his best to summon up an answer. " 'Too ill success will be dangerous'—let them fear that who allow excuses, or can be content to overlive their honour. 'Too great will be envious'—I will never foreswear virtue for fear of ostracism. 'The Court is the centre.'—But methinks it is the fairer choice to command armies than humours." ... "These are the very private problems," he concluded, "and nightly disputations, which from your Lordship, whom I account another myself, I cannot hide."

At moments the gloom lifted, and hope returned. The Queen smiled; disagreements vanished; something like the old happy confidence was in the air once more. On Twelfth Night, 1599, there was a grand party for the Danish ambassador, and the Queen and the Earl danced hand in hand before the assembled Court. Visions of that other Twelfth Night, five short years before—that apogee of happiness—must have flitted through many memories. Five short years—what a crowded gulf between then and now! And yet, now as then, those two figures were together in their passion and their mystery, while the viols played their beautiful tunes and the jewels glittered in the torch-light. What was passing? Perhaps, in that strange companionship, there was delight, as of old . . . and for the last time.

Elizabeth had much to trouble her—Ireland, Essex, the eternal question of War and Peace—but she brushed it all aside, and sat for hours translating the *Ars Poetica* into English prose. As for Ireland, she had grown accustomed to that; and Essex, though fretful, seemed only anxious to cut a figure as Lord Deputy—she could ignore those uncomfortable suspicions of a few months ago. There remained the Spanish War; but that too seemed to have solved itself very satisfactorily. It drifted on, in complete ambiguity, while peace was indefinitely talked of, with no fighting and no expense; a war that was no war, in fact—precisely what was most to her liking.

One day, however, she had a shock. A book fell into her hands—a History of Henry the Fourth—she looked at it—there was a Latin dedication to Essex.

"To the most illustrious and honoured Robert Earl of Essex and Ewe, Earl Marshal of England, Viscount of Hereford and Bourchier, Baron Ferrars of Chartley, Lord Bourchier and Louen"—what was all this? She glanced through the volume, and found that it contained an elaborate account of the defeat and deposition of Richard the Second—a subject, implying as it did the possibility of the removal of a sovereign from the throne of England, to which she particularly objected. It was true, no doubt, that the Bishop of Carlisle was made to deliver an elaborate speech against the King's deposition; but why bring the matter before the public at all? What could be the purpose of this wretched book? She looked again at the dedication, and as she looked the blood rushed to her head. The tone was one of gross adulation, but that was by no means all; there was a phrase, upon which a most disgraceful construction might be put. "Most illustrious Earl, with your name adorning the front of our Henry, he may go forth to the public happier and safer."[1] The man would, no doubt, pretend that "our Henry" referred to the book; but was there not another very possible interpretation?—that if Henry IV had possessed the name and titles of Essex his right to the throne would have been better and more generally recognised. It was treason! She

[1] Illustrissime comes, cujus nomen si Henrici nostri fronti radiaret, ipse et laetior et tutior in vulgus prodiret.

sent for Francis Bacon. "Cannot this man—this John Hayward—be prosecuted for treason?" she asked. "Not, I think, for treason, Madam," was the reply, "but for felony." "How so?" "He has stolen so many passages from Tacitus. . . ." "I suspect the worst. I shall force the truth from him. The rack——." Bacon did what he could to calm her; but she was only partially pacified; and the unfortunate Hayward, though he was spared the rack, was sent to the Tower, where he remained for the rest of the reign.

Her suspicions, having flamed up in this unexpected manner, sank down again, and, after a slight scene with Essex, she finally signed his appointment as Lord Deputy. He departed at the end of March, passing through the streets of London amid the acclamations of the citizens. In the popular expectation, all would be well in Ireland, now that the Protestant Earl had gone there to put things to rights. But, at Court, there were those whose view of the future was different. Among them was Bacon. He had followed the fluctuations of the Irish appointment with interest and astonishment. Was it really possible that, with his eyes open, that rash man had fallen into such a trap? When he found that it was indeed the case, and that Essex was actually going, he wrote him a quiet, encouraging letter, giving no expression to his fears or his doubts. There was nothing else to be done; the very intensity of his private conviction made a warning useless and impossible. "I did as plainly see," he afterwards wrote, "his overthrow chained, as it were, by destiny to that journey as it is possible for a man to ground a judgment upon future contingents."

XII

THE state of affairs in Ireland was not quite so bad as it might have been. After the disaster on the Blackwater, rebellion had sprung up sporadically all over the island; the outlying regions were everywhere in open revolt; but Tyrone had not made the most of his opportunity, had not advanced on Dublin, but had frittered away the months during which he had been left undisturbed by his enemies in idleness and indecision. He was a man who was more proficient in the dilatory arts of negotiation—sly bargaining, prolonged manoeuvring, the judicious making and breaking of promises—than in the vigorous activities of war. Of Irish birth and English breeding, half savage and half gentleman, half Catholic and half sceptic, a schemer, a lounger, an adventurer, and a visionary, he had come at last, somehow or other, after years of diffused cunning, to be the leader of a nation and one of the pivots upon which the politics of Europe turned. A quiet life was what he longed for—so he declared; a quiet life, free alike from the intolerance of Protestantism and the barbarism of war; and a quiet life, curiously enough, was what in the end he was to be given. But the end was not yet, and in the meantime all was disturbance and uncertainty. It had been impossible for him to assimilate his English Earldom with the chieftainship of the O'Neils. His hesitating attempts to be a loyal vassal of the Saxons had yielded to the pressure of local patriotism; he had intrigued and rebelled; he had become the client of Philip of Spain. More than once the English had held him at their mercy, had accepted his submission, and had reinstated him in his honours and his lands. More than once, after trading on their fluctuating policies of severity and moderation, he had treacherously turned against them the power and the influence which their protection had enabled him to acquire. Personal animosities had been added to public feuds. He had seduced the sister of Sir Henry Bagenal, had carried her off and married her, in spite of her brother's teeth; she had died in misery; and Sir Henry, advancing with his army to meet the rebel at the Blackwater, had been defeated and killed. After such a catastrophe, it seemed certain that the only possible issue was an extreme one. This time the English Government would admit no compromise, and Tyrone must be finally crushed. But Tyrone's own view was very different; he was averse from extremity; he lingered vaguely in Ulster; the old system of resistance, bargaining, compromise, submission, and reconciliation, which had served him so often, might very well prove useful once again.

But one thing was clear: if the English Government desired the speedy destruction of Tyrone, it could have chosen no one more anxious to second its purposes than the new Lord Deputy. For Essex, it was obvious, an Irish victory was vital. Would he achieve one? Francis Bacon was not the only observer at Court to be pessimistic on that subject. A foreboding gloom was in the air. When John Harington was about to follow his patron to Ireland with a command in the Cavalry, he received from his kinsman, Robert Markham, who had an office about the Court, a weighty letter of advice and instruction. Harington was bidden to be most careful in his conduct; there would be spies in the Irish army, who would report everything to high-placed ill-wishers at home. "Obey the Lord Deputy in all things," wrote Markham, "but give not your opinion; it may be heard in England." The general situation, Markham thought, was menacing. "Observe," he said, "the man who commandeth, and yet is commanded himself; he goeth not forth to serve the Queen's realm, but to humour his own revenge" . . . "If the Lord Deputy," he went on, "performs in the field what he hath promised in the Council, all will be well; but, though the Queen hath granted forgiveness for his late demeanour in her presence, we know not what to think hereof. She hath, in all outward semblance, placed confidence in the man who so lately sought other treatment at her hands; we do sometime think one way, and sometime another; what betideth the Lord Deputy is known to Him only who knoweth all; but when a man hath so many shewing friends and so many unshewing enemies, who learneth his end below? . . . Sir William Knollys is not well pleased, the Queen is not well pleased, the Lord Deputy may be pleased now, but I sore fear what may happen hereafter."

To such warnings, no doubt, Harington—a gay spark, who had translated Ariosto into English verse and written a Rabelaisian panegyric on water closets—paid no great heed; but in fact they expressed, with an exactness that was prophetic, the gist of the situation. The expedition was a gamble. If Essex won in Ireland, he won in England, too; but the dice were loaded against him; and if he failed . . . From the very first, the signs were unpropitious. The force of sixteen thousand foot and fifteen hundred horse, which had been collected for the expedition, was, for an Elizabethan army, a well-equipped and efficient one; but that was the beginning and the end of the Lord Deputy's advantages. His relations with the Home Government were far from satisfactory. Elizabeth distrusted him—distrusted his capacity and even, perhaps, his intentions; and the Secretary, who now dominated the Council, was his rival, if not his enemy. His wishes were constantly thwarted, and his decisions over-ruled. A serious quarrel broke out before he had left England. He had appointed Sir Christopher Blount to be one of his Council, and Lord Southampton his General of the horse; both appointments were cancelled by Elizabeth. Her objections to Sir Christopher are unknown—possibly she considered his Catholicism a bar to high position in Ireland; but Southampton, who had incurred her supreme displeasure by carrying on an intrigue with Elizabeth Vernon, one of her ladies-in-waiting, and then daring to marry her—Southampton, whom, in her fury, she had put into prison together with his bride—that Essex should have ventured to name this young reprobate for a high command seemed to her little short of a deliberate impertinence. There was some fierce correspondence; but she held firm; the two men followed Essex as private friends only; and the Lord Deputy arrived in Dublin—it was April 1599—in a gloomy mood and a fretted temper.

He was immediately faced with a strategical question of crucial importance. Should he at once proceed to Ulster and dispose of Tyrone, or should he first suppress the smouldering disaffection in the other parts of the island? The English Council in Dublin recommended the latter course, and Essex agreed with them. It would be easier, he thought, to deal with the main forces of the rebellion when its subsidiary supports had been demolished. Possibly he was right; but the decision implied a swift and determined execution; to waste too much time and too much energy on minor operations would be worse than useless. That was obvious, and the subduing of a few recalcitrant chiefs with a powerful English Army seemed a simple enough affair. Essex marched into Leinster, confident that nothing could resist him —and nothing could. But he was encountered by something more dangerous than resistance—by the soft, insidious, undermining atmosphere of that paradoxical country which, a quarter of a century earlier, had brought his father to despair and death.

The strange air engulfed him. The strange land—charming, savage, mythical—lured him on with indulgent ease. He moved, triumphant, through a new peculiar universe of the unimagined and the unreal. Who or what were these people, with their mantles and their nakedness, their long locks of hair hanging over their faces, their wild battle-cries and gruesome wailings, their kerns and their gallowglass, their jesters and their bards? Who were their ancestors? Scythians? Or Spaniards? Or Gauls? What state of society was this, where chiefs jostled with gypsies, where ragged women lay all day long laughing in the hedgerows, where ragged men gambled away among each other their very rags, their very forelocks, the very . . . parts more precious still, where wizards flew on whirlwinds, and rats were rhymed into dissolution? All was vague, contradictory, and unaccountable; and the Lord Deputy, advancing further and further into the green wilderness, began—like so many others before and after him—to catch the surrounding infection, to lose the solid sense of things, and to grow confused over what was fancy and what was fact.

His conquering army was welcomed everywhere by the English settlers. The towns threw open their gates to him, and he was harangued in Latin by delighted Mayors. He passed from Leinster into Munster—still victorious. But time was slipping away. Days and days were spent over the reduction of unimportant castles. Essex had never shown any military genius—only a military taste; and his taste was gratified now, as it had never been before, by successful skirmishes, romantic escapades, noble gestures, and personal glory. The cost was serious. He had lost sight of his main purpose in a tangle of insignificant incidents. And while he was playing with time, his strength was dwindling. Under the combined influences of casualties, desertions, disease, and the garrisoning of distant outposts, his army was melting away. At last, in July, he found himself back in Dublin, having spent nearly three months in dubious operations far from the real force of the enemy, and with the numbers of the men under his command diminished by one half.

Then the mist of illusion melted, and he was faced

with the deplorable truth. At this late hour, with his weakened army, was it possible any longer to make sure of crushing Tyrone? In extreme agitation he counted up the chances, and knew not which way to turn. Wherever he looked, a gulf seemed to open at his feet. If he failed against Tyrone, how fatal! If he did nothing, what a derision! Unable to bring himself to admit that he had muddled away his opportunity, he sought relief in random rage and wild accusations, in fits of miserable despair, and passionate letters to Elizabeth. A detachment of some hundreds of men had shown cowardice in the field; he cashiered and imprisoned all the officers, he executed a lieutenant, and he had every tenth man in the rank and file put to death. He fell ill, and death seemed to come near to him too; he would welcome it. He rose from his couch to write a long letter to the Queen, of exposition and expostulation. "But why do I talk of victory or success? Is it not known that from England I receive nothing but discomfort and soul's wounds? Is it not spoken in the army, that your Majesty's favour is diverted from me, and that already you do bode ill both to me and it? . . . Is it not lamented of your Majesty's faithfullest subjects, both there and here, that a Cobham or a Raleigh—I will forbear others for their places' sakes—should have such credit and favour with your Majesty when they wish the ill-success of your Majesty's most important action? . . .

Let me honestly and zealously end a wearisome life. Let others live in deceitful and inconstant pleasures. Let me bear the brunt, and die meritoriously. . . . Till then, I protest before God and His Angels, I am a true votary, that is sequestered from all things but my duty and my charge. . . . This is the hand of him that did live your dearest, and will die your Majesty's faithfullest servant."

There was a sudden rising in Connaught which had to be put down; the rebels were defeated by Sir Christopher Blount; but by now July was over, and the Lord Deputy was still in Dublin. Meanwhile, at home, as time flowed by, and no news of any decisive action came from Ireland, men's minds were divided between doubt and expectation. At Court, the tone was cynical. "Men marvel," a gossip wrote on August 1, "Essex hath done so little; he tarries yet at Dublin." The decimation of the soldiers was "not greatly liked," and when news came that the Lord Deputy had used the powers specially given him by the Queen to make no fewer than fifty-nine knights, there was much laughter and shrugging of shoulders. But elsewhere the feeling was different. The people of London still had high hopes for their favourite—hopes which were voiced by Shakespeare in a play which he produced at this moment at the Globe Theatre. Southampton was the friend and patron of the rising dramatist, who took this opportunity of making a graceful public allusion to Southampton's own patron and friend.

"How London doth pour out her citizens!"

So spoke the Chorus in "Henry V," describing the victorious return of the King from France—

> "As, by a lower but by loving likelihood,
> Were now the general of our gracious Empress,
> As in good time he may, from Ireland coming,
> Bringing rebellion broachèd on his sword,
> How many would the peaceful city quit
> To welcome him!"

The passage was no doubt applauded, and yet it is possible to perceive even here, through the swelling optimism of the lines, a trace or two of uneasiness.

Elizabeth, waiting anxiously for a despatch announcing Tyrone's defeat, and receiving instead nothing but letter after letter of angry complaints and despairing ejaculations, began to grow impatient. She did not restrain her comments to those about her. She liked nothing, she said, that was done in Ireland. "I give the Lord Deputy a thousand pounds a day to go on progress." She wrote to him complaining bitterly of the delay, and ordering him to march forthwith into Ulster. The reply came that the army was fatally depleted—that only 4000 men were left of the 16,000 that had gone from England. She sent a reinforcement of 2000; but the expense cut her to the quick. What was the meaning of this waste and this procrastination? Sinister thoughts came floating back into her head. Why, for instance, had he made so many knights? She wrote, peremptorily ordering Essex to attack Tyrone, and not to leave Ireland till he had done so. "After you shall have certified us to what form you have reduced things in the North . . . you shall with all speed receive our warrant, without which we do charge you, as you tender our pleasure, that you adventure not to come out of that kingdom by virtue of any former license whatever."

Her agitation deepened. One day at Nonesuch she met Francis Bacon, and drew him aside. She knew him as a clever man, a friend of Essex, and possibly she could extract something from him which would throw a light on the situation. What was his opinion, she asked, of the state of affairs in Ireland, and—she flashed a searching glance on him—the proceedings of the Lord Deputy? It was an exciting moment for Bacon. The honour was great and unexpected—he felt himself swept upward. With no official standing whatever, he was being consulted in this highly confidential way. What was he to answer? He knew all the gossip, and had reason to believe that, in the Queen's opinion, Essex was acting in a manner that was not only unfortunate and without judgment, but "contemptuous and not without some private end of his own." With this knowledge, he made a reply that was remarkable. "Madam," he said, "if you had my Lord of Essex here with a white staff in his hand, as my Lord of Leicester had, and continued him still about you for society to yourself, and for an honour and ornament to your attendance and Court in the eyes of your people, and in the eyes of foreign ambassadors, then were he in his right element. For to discontent him as you do, and yet to put arms and power into his hands, may be a kind of temptation to make him prove cumbersome and unruly. And therefore if you would send for him, and satisfy him with honour here near you, if your affairs—which I am not acquainted with—will permit it, I think were the best way." She thanked him, and passed onwards. So that was how the land lay! "Arms and power . . . temptation . . . cumbersome and unruly!" He had blown upon her smouldering suspicions, and now they were red hot.

Shortly afterwards Henry Cuffe arrived from Ireland, with letters and messages for the Queen from the Lord Deputy. The tale he had to tell was by no means reassuring. The army, weakened still further by disease and desertion, was in an unsatisfactory condition; the bad weather made movement difficult; and the Dublin Council had once more pronounced strongly against an attack upon Ulster. Elizabeth wrote a scathing letter to her "right trusty and well beloved cousin," in which she no longer gave command, but merely desired to be informed what he was going to do next. She could not imagine, she said, what could be the explanation of his conduct. Why had nothing been done? "If sickness of the army be the reason, why was not the action undertaken when the army was in better state? If winter's approach, why were the summer months of July and August lost? If the spring were too soon, and the summer that followed otherwise spent, and the harvest that succeeded were so neglected as nothing hath been done, then surely we must conclude that none of the four quarters of the year will be in season for you and that Council to agree to Tyrone's prosecution, for which all our charge is intended." Then, into the middle of her long and bitter argumentation, she stuck a phrase well calculated to give a jar to her correspondent. "We require you to consider whether we have not great cause to think that your purpose is not to end the war." She was determined to make him realise that she was watching him carefully and was prepared for any eventuality.

Meanwhile, in Dublin, the moment of final decision was swiftly approaching. The horns of a fearful dilemma were closing in upon the unfortunate Lord Deputy. Was he to obey the Queen, and risk all against his own judgment and the advice of his Council? Or was he to disobey her, and confess himself a failure? Winter was at hand, and, if he were going to fight, he must fight at once. Hysterical and distracted, he was still hesitating, when letters were brought to him from England. They told him that Robert Cecil had been appointed to the lucrative office, which he himself had hoped to receive, of the Mastership of the Wards. Then every other feeling was drowned in rage. He rushed to Blount and Southampton. He had made up his mind, he said; he would not go into Ulster; he would go into England, at the head of his army; he would assert his power; he would remove Cecil and his partners; and he would make sure that henceforward the Queen should act as she ought to act and as he wished.

The desperate words were spoken, but that was all. The hectic vision faded, and, before the consultation was over, calmer councils had prevailed. Sir Christopher pointed out that what the Earl was proposing—to lead his small army, with such a purpose, from Wales to London—meant civil war. It would be wiser, he said, to go over with a bodyguard of a few hundred tried followers, and effect a *coup d'état* at Nonesuch. But this plan too was waved aside. Suddenly veering, Essex decided to carry out the Queen's instructions and to attack Tyrone in Ulster.

As a preliminary, he ordered Sir Conyers Clifford, at the head of a picked force, to effect a diversion by marching against the rebels from Connaught. He himself was preparing to move, when there was a new catastrophe: Clifford, caught by the enemy on a causeway crossing a bog, was set upon, defeated, and killed. But it was too late for Essex to draw back, and at the end of August he left Dublin.

At the same time he composed and despatched a short letter to the Queen. Never were his words more gorgeous and his rhythms more moving: never were the notes of anguish, remonstrance, and devotion so romantically blended together.

"From a mind delighting in sorrow; from spirits wasted with travail, care, and grief; from a heart torn in pieces with passion; from a man that hates himself and all things that keep him alive, what service can your Majesty reap? Since my services past deserve no more than banishment and proscription into the most cursed of all countries, with what expectation and what end shall I live longer? No, no, the rebel's pride and successes must give me means to ransom myself, my soul I mean, out of this hateful prison of my body. And if it happen so, your Majesty may believe that you shall not have cause to mislike the fashion of my death, though the course of my life may not please you. From your Majesty's exiled servant, ESSEX."

It was very fine—thrilling, adorable! But the sequel was less so. If the desperate knight had indeed flung himself to death amid the arrows of the barbarians . . . but what happened was altogether different. In a few days he was in touch with Tyrone's army, which, though it outnumbered his own, refused to give battle.

There was some manoeuvring, a skirmish, and then Tyrone sent a messenger, demanding a parley. Essex agreed. The two men met alone, on horseback, at a ford in a river, while the armies watched from either bank. Tyrone, repeating his old tactics, offered terms —but only verbally; he preferred, he said, not to commit them to writing. He proposed a truce, to be concluded for six weeks, to continue by periods of six weeks until May Day, and not to be broken without a fortnight's warning. Essex again agreed. All was over. The campaign was at an end.

Of all possible conclusions, this surely was the most impotent that could have been imagined. The grand expedition, the noble general, efforts, hopes, vaunting —it had all dwindled down at last to a futile humiliation, an indefinite suspension of hostilities—the equivocal, accustomed triumph of Tyrone. Essex had played all his cards now—played them as badly as possible, and there was nothing left in his hand. Inevitably, as the misery of his achievement sank into his consciousness, the mood of desperate resolutions returned. He decided that there was only one thing now that could save the situation—he must see the Queen. But—such was the wild wavering of his spirit —whether he was to come into her presence as a suppliant or as a master, he could not tell: he only knew that he could bear to be in Ireland no longer. With Blount's suggestion of a *coup d'état* indeterminately hovering in his mind, he summoned round him the members of his household, and, accompanied by them and a great number of officers and gentlemen, embarked at Dublin on September 24th. Early on the morning of the 28th the troop was galloping into London.

The Court was still at Nonesuch, in Surrey, about ten miles southward; the river lay between; and, if an attack were to be made, it would be necessary for the cavalcade to ride through the City and cross the Thames at London Bridge. But by this time the notion of deliberate violence had become an unreality —had given place to the one overmastering desire to be with the Queen at the earliest possible moment. The quickest way was to take the ferry from Westminster to Lambeth, and Essex, leaving the bulk of his followers to disperse themselves in London, had himself rowed across the river with six of his chosen friends. At Lambeth the weary men seized what horses they could find and rode on. They were soon passed by Lord Grey of Wilton, a member of the Cecil party, who, on a fresher mount, was also riding to Court that morning. Sir Thomas Gerard spurred after him. "My Lord, I beg you will speak with the Earl." "No," was Lord Grey's reply, "I have business at Court." "Then I pray you," said Sir Thomas, "let my Lord of Essex ride before, that he may bring the first news of his return himself." "Doth he desire it?"

said Lord Grey. "No," said Sir Thomas, "nor I think will desire anything at your hands." "Then I have business," said Lord Grey, and rode on with greater speed than ever. When Gerard told his friends what had occurred, Sir Christopher St. Lawrence cried out with an oath that he would press on and kill Lord Grey, and after him the Secretary. The possibility of a swift, dramatic, irretrievable solution hovered in the air for a moment amid the group of angry gentlemen. But Essex forbade it; it would be mere assassination; he must take his chance.

Directly Lord Grey reached Nonesuch he went to Cecil, and told him the astounding news. The Secretary was calm; he did nothing—sent no word to the Queen, who was dressing in her upper chamber—but waited quietly in his chair. A quarter of an hour later—it was ten o'clock—the Earl was at the gate. He hurried forward, without a second's hesitation; he ran up the stairs, and so—oh! he knew the way well enough—into the presence chamber, and thence into the privy chamber; the Queen's bedroom lay beyond. He was muddy and disordered from his long journey, in rough clothes and riding boots; but he was utterly unaware of any of that, as he burst open the door in front of him. And there, quite close to him, was Elizabeth among her ladies, in a dressing-gown, unpainted, without her wig, her grey hair hanging in wisps about her face, and her eyes starting from her head.

XIII

SHE was surprised, she was delighted—those were her immediate reactions; but then, swiftly, a third feeling came upon her—she was afraid. What was the meaning of this unannounced, this forbidden return, and this extraordinary irruption? What kind of following had the man brought from Ireland and where was it? What had happened? Was it possible that at this very moment she was in his power? Completely in the dark, she at once sought refuge in the dissimulation which was her second nature. Her instinctive pleasure in his presence, her genuine admiration of his manner and his speech, served her purpose excellently, and, covered with smiles, she listened while he poured out his protestations and told his story—listened with an inward accompaniment of lightning calculations and weighings of shifting possibilities and snatchings at dubious hints. Very soon she guessed that she was in no immediate danger. She laughingly bade him begone and change his clothes, while she finished her toilet; he obeyed, returned, and the conversation continued for an hour and a half. He came downstairs to dinner in high spirits, flirted with the ladies, and thanked God that after so many storms abroad he had found so sweet a calm at home. But the calm was of short continuance; he saw the Queen again after dinner and found the breezes blowing. She had made her inquiries, and, having sufficiently gauged the situation, had decided on her course of action. She began by asking disagreeable questions, disagreeably; when he answered, she grew angry; finally she declared that he must explain himself to the Council. The Council met, and when the Earl had given an account of his proceedings, adjourned in vague politeness. Perhaps all was well—it almost seemed so; but the Queen, apparently, was still vexed and inaccessible. At eleven o'clock at night the Earl received a message from Her Majesty; he was commanded to keep to his chamber.

Every one was mystified, and the wildest speculations flew about. At the first blush it was supposed that Essex had completely triumphed—that in one bold stroke he had recaptured the favour and the power that were slipping from his grasp. Bacon sent off a letter of congratulation. "I am more yours than any man's and more yours than any man," he wrote. A little later, the news of the Queen's displeasure brought doubts; yet it seemed hardly possible that anything very serious should happen to the Earl, who, after all, had only been blundering in Ireland, like so many before him. But meanwhile the Queen proceeded with her plan. Having waited a day, during which no news came of any suspicious movements in London, she felt she could take her next step. She committed Essex to the custody of the Lord Keeper Egerton, to whose residence—York House, in the Strand—he was forthwith removed. All still remained calm, and Elizabeth was satisfied: Essex was now completely at her mercy. She could decide at her leisure what she would do with him.

While she was considering he fell ill. He had been seriously unwell before he left Ireland, and the fatigue of his three days' ride across England, followed by the emotion and disgrace at Nonesuch, had proved too much for his uncertain and suggestible physique. Yet, while he lay in captivity at York House, he still —though crying out from time to time that he only longed for a country obscurity—had not given up hopes of a return to favour and even a reinstatement as Lord Deputy. He wrote submissive letters to the Queen; but she refused to receive them, and sent no word. John Harington, who had been among those he had knighted in Ireland, returned at this moment, and Essex begged him to be the bearer of yet another missive, filled with contrition and adoration. But the sprightly knight preferred to take no risks. He had been threatened with arrest on his arrival in London, and he felt that his own affairs were as much as he could manage; charity, he said, began at home, and he had no desire to be "wracked on the Essex coast."

His conscience, too, was not quite clear. He had had the curiosity to pay a visit to Tyrone after the pacification, and had behaved, perhaps, in too friendly and familiar a fashion with the recreant Earl. He had produced a copy of his Ariosto, had read aloud some favourite passages, had presented the book to the elder of Tyrone's sons—"two children of good towardly spirit, in English clothes like a nobleman's sons, with velvet jerkins and gold lace,"—and finally had sat down to a merry dinner with the rebels at a "fern table, spread under the stately canopy of heaven." Possibly some rumour of these proceedings had reached Elizabeth's ears, and she was not altogether pleased by them. Nevertheless he believed that all would be well if only he could obtain an audience. He knew that she had a liking for him; he was her godson—had been familiar with her from his childhood, and was actually connected, in an underground way, with the royal family, his stepmother having been a natural daughter of Henry VIII. At last he was told that the Queen would receive him; he went to Court in considerable trepidation; and as soon as he entered the presence he thanked his stars that he had had the sense to refuse to deliver any message from Essex.

He never forgot the fearful scene that followed. Hardly had he knelt before her than she strode towards him, seized him by the girdle, and, shaking it, exclaimed, "By God's Son, I am no Queen! That man is above me! Who gave him command to come here so soon? I did send him on other business." While the terrified poet stammered out some kind of answer, she turned from him in fury, "walked fastly to and fro," and "looked with discomposure in her visage." "By God's Son!" she burst out again, "you are all idle knaves and Essex worse!" He tried to pacify her, but "her choler did outrun all reason," she would listen to nothing, and, in the storm of her invective, seemed to forget that her unfortunate godson was not, after all, the Lord Deputy. At last, however, she grew calmer, asked questions, was amused by Harington's little jokes and stories, and made no account of his hobnobbing with Tyrone. He described the rebel to her, and his curious Court—how "his guard for the most part were beardless boys, without shirts, who in the frost wade as familiarly through rivers as water-spaniels." "With what charm," he added, "such a master makes them love him, I know not; but if he bid them come, they come; if go they do go; if he say do this, they do it." She smiled; and then, suddenly changing countenance, told him to go home. He "did not stay to be bidden twice," but rode away to his house in Somersetshire "as if all the Irish rebels had been at his heels."

The author of the *Metamorphosis of Ajax* was no fit confidant for a perplexed and injured sovereign.

Elizabeth looked elsewhere for an adviser, or at any rate a listener, and she found what she wanted in Francis Bacon. Recalling the conversation of the summer, she took advantage of his official attendance upon her on legal business to revert to the subject of the Earl. She found his answers pertinent; she renewed the topic; and so began a series of strange dialogues in which, during many months, in confidential privacy, the fate of Essex, with all its hidden implications of policy and passion, became the meeting-point of those two most peculiar minds. Elizabeth was, as usual, uncertain how to treat the situation in which she found herself: was there to be forgiveness or punishment? and, if the latter, of what kind? Revealing little, she asked much. As for Bacon, he was in his element. He felt that he could thread his way through the intricacies that surrounded him with perfect propriety. To adjust the claims of personal indebtedness and public duty, to combine the feelings of the statesman and the friend, to hold the balance true between honour and ambition—other men might find such problems difficult, if not insoluble; but he was not frightened by them; his intellect was capable of more than that. As he talked to Elizabeth, he played upon the complex theme with the profound relish of a virtuoso. He had long since decided that, in all human probability, Essex was a ruined man; he owed the Earl something—much; but it would be futile to spoil his own chances of fortune by adhering to a hopeless cause; it was essential to win the good graces of Robert Cecil; and now, there was this heaven-sent opportunity—which it would be madness to miss—for acquiring something more important still—the confidence of the Queen. Besides—he could doubt it no longer—Essex was a mischievous person, whose activities were dangerous to the State. While he was clearly bound to give him what help he could as a private individual, he was certainly under no obligation to forward the return of such a man to power; it was even his duty to insinuate into the Queen's mind his own sense of the gravity of the situation. And so, with unhesitating subtlety, he spun the web of his sagacious thought. He had no doubt of himself—none; and when, a few years later, under the pressure of the public disapproval, he wrote an account of his proceedings, it still seemed to him that a recital of his actual conduct was all that was necessary as a justification.

Elizabeth listened with interest to everything he had to say—it was always impossible to do otherwise. He was profuse in his expressions of sympathy and attachment to the Earl; but, he must needs say it, there were some positions to which he thought him ill-suited; to send him back to Ireland, for instance— "Essex!" interrupted the Queen. "Whensoever I send Essex back again into Ireland, I will marry you.

Claim it of me." No, that was not her thought—far from it; she intended rather to bring him to justice; but by what process? She inclined to a trial before the Star Chamber. But Bacon demurred. It would, he said, be a dangerous proceeding; it might be difficult to produce cogent proof in public of the Earl's delinquencies; and his popularity was so great that a severe punishment on insufficient evidence might produce most serious consequences. She glared angrily, and dismissed him. She did not like that suggestion; but the words sank into her mind, and she veered away from the notion of a public prosecution.

For, as time passed, everything seemed to show that Bacon's warning was justified. There could be no doubt about the Earl's popularity. It was increased by his illness, and, when it was whispered that he lay near to death in his captivity, the public indignation made itself heard. Pamphlets, defending the Earl and attacking his enemies, were secretly printed and scattered broadcast. At last, even the white walls of the palace were covered with abusive scrawls. Bacon was singled out for particular denunciation; he was a traitor, who was poisoning the Queen's mind against his benefactor. He was threatened—so he declared—with assassination. This was unpleasant, but some use might be made of it: it might serve to put beyond a doubt his allegiance to the Secretary. He wrote to his cousin, telling him of these threats of violence, against which, he said, "I thank God I have the privy coat of a good conscience." He looked upon them "as a deep malice to your honourable self, upon whom, by me, through nearness, they think to make some aspersion."[2]

Cecil smiled gently when he read the letter; and he sent for his cousin. He wished to make his own position quite clear. He had indeed heard, he said, that Francis had been doing some ill office to Essex; but . . . he did not believe it. And then he added: "For my part, I am merely passive and not active in this action; and I follow the Queen, and that heavily, and I lead her not . . . The Queen indeed is my Sovereign, and I am her creature, I may not leese her; and the same course I wish you to take."

So he explained himself, and the explanation was a perfectly true one. Robert Cecil was indeed merely passive, merely following, with the sadness which his experience of the world had brought him, the action of the Queen. But passivity, too, may be a kind of action—may, in fact, at moments prove more full of consequence than action itself. Only a still, disillusioned man could understand this; it was hidden from the hasty children of vigour and hope. It was hidden, among others, from Walter Raleigh. He could not conceive what the Secretary was doing; he was letting a golden opportunity slip through his fingers; he was leaving the Queen to her own devices—it was madness—this was the time to strike. "I am not wise enough," he wrote to Cecil, "to give you advice; but if you take it for a good counsel to relent towards this tyrant, you will repent it when it shall be too late. His malice is fixed, and will not evaporate by any your mild courses. For he will ascribe the alteration to Her Majesty's pusillanimity and not to your good nature: knowing that you work but upon her humour, and not out of any love towards him. The less you make him, the less he shall be able to harm you and yours. And if Her Majesty's favour fail him, he will again decline to a common person. . . . Lose not your advantage; if you do, I rede your destiny. Yours to the end, W. R."[3] It was true—he was not "wise enough" to give a Cecil advice. Could he not see that the faintest movement, the slightest attempt to put pressure upon the Queen, would be fatal? How little he understood that perverse, that labyrinthine character! No! If anything was to be done, she herself, in her own strange way and with her own strange will, must do it. And the Secretary sat motionless—waiting, watching, and holding his breath.

Elizabeth, certainly, needed watching very carefully. For the moment she seemed to be occupied with entirely frivolous pursuits. The ceremonies of Accession Day absorbed her; she sat for hours in the tilt-yard—where Essex had so often shone in all his glory—careless and amused; and when at last there was a grotesque surprise and Lord Compton came in, as an eye-witness described it, "like a Fisherman, with 6 men clad in motley, his capariesons all of nett, having caught a Frogge," the old creature's sides shook with delighted laughter. A week later she came to a sudden decision: she would justify her treatment of Essex before the world by having a statement of his delinquencies read out by the Council in the Star Chamber. He himself could not be present—he was too ill. But was he? She could not feel quite sure; he had been known before now to convert a fit of the sulks into a useful malady; she would see for herself. And so, at four o'clock in the evening of November 28th, accompanied by Lady Warwick and Lord Worcester, she stepped into her barge and had herself conveyed to York House. We know no more. Essex was in truth very ill—apparently dying. Was he conscious of her visit? Were there words spoken? Or did she come and look and go, unseen? Unanswerable questions! The November night falls, gathering her up into its darkness.

Next day the Star Chamber met, and the statement of the Earl's misdoings was read aloud. It was declared that he had mismanaged the Irish operations, that he had made a disgraceful treaty with Tyrone, and that he had returned to England contrary to the Queen's express orders. Members of the public were admitted, but Francis Bacon did not attend. Elizabeth, running over the list of those who had been present, observed the fact. She sent him a message, asking the meaning of it. He replied that he had thought it wiser to keep away, in view of the threats of violence against his person. But she was not impressed by the excuse, and did not speak to him again for several weeks.

The Star Chamber declaration led to nothing. The weeks, the months, flowed by, and Essex was still a prisoner; the fatal evening at Nonesuch proved to have been the beginning of a captivity which lasted almost a year. Nor was it a mild one. None of the Earl's intimates were allowed to see him. Even Lady Essex, who had just borne him a daughter, and who haunted the Court dressed in the deep mourning of a suppliant, was forbidden to see her husband for many months. Elizabeth's anger had assumed a grimmer aspect than ever before. Was this still a lovers' quarrel? If so, it was indeed a strange one. For now contempt, fear, and hatred had come to drop their venom into the deadly brew of disappointed passion. With fixed resentment, as the long months dragged out, she nursed her wrath; she would make him suffer for his incompetence, his insolence, his disobedience; did he imagine that his charms were irresistible? She had had enough of them, and he would find that he had made a mistake.

With the new year—it was the last of the century—there were two developments. Essex began to recover, and by the end of January he had regained his normal health. At the same time the Queen made a new attempt to deal with the situation in Ireland. Tyrone had himself put an end to the truce of September, and had recommenced his manoeuvrings against the English. Something had to be done, and Elizabeth, falling back on her previous choice, appointed Mountjoy Lord Deputy. He tried in vain to escape from the odious office, but it was useless; Elizabeth was determined; go he must. Before doing so, however, he held a consultation with Southampton and Sir Charles Davers, another devoted follower of Essex, as to how he might best assist the imprisoned Earl. An extraordinary proposal was made. For some years past Essex had been in communication with James of Scotland, and Mountjoy himself, during the campaign in Ireland, had written to the King—whether with or without the knowledge of Essex is uncertain—asking him to make some move in Essex's behalf. James's answer having proved unsatisfactory, the matter was dropped; but it was now revived in an astonishing and far more definite manner. It was well known that the prime object of the King of Scotland's policy was to secure the inheritance of England. Mountjoy suggested that a message should be sent to James informing him that the Cecil party was hostile to his succession, that his one chance lay in the reinstatement of Essex, that if he would take action in Essex's favour Mountjoy himself would cross over from Ireland with an army of four or five thousand men, and that with their combined forces they could then impose their will upon the English Government. Southampton and Davers approved of the project, and there can be no doubt that Essex himself gave his consent to it, for the conspirators had found means of conveying letters in secret to and from York House. The messenger was despatched to Scotland; and Mountjoy actually started to take up the government of Ireland with this project of desperate treason in his mind. But James was a cautious person: his reply was vague and temporising; Mountjoy was informed; and the scheme was allowed to drop.

But not for long. For in the spring, Southampton went to Ireland, and Essex took the opportunity to send a letter to Mountjoy, urging him to carry out his original intention and to lead his army into England, with or without the support of James. Mountjoy, however, had changed his mind. Ireland had had its effect on him too—and an unexpected one. He was no longer the old Charles Blount, who had been content to follow in the footsteps of his dazzling friend; he had suddenly found his vocation. He was a follower no more; he was a commander; he felt that he could achieve what no one had achieved before him; he would pacify Ireland, he would defeat

Tyrone. Penelope herself would not keep him from that destiny. His answer was polite, but firm. "To satisfy my lord of Essex's private ambition, he would not enter into an enterprise of that nature."[1]

Meanwhile Elizabeth, unaware of these machinations, was wondering gloomily what she was going to do. The Tower? On the whole, she thought not; things were bad—but not quite so bad as that. Nevertheless, she would move the culprit out of York House. The poor Lord Keeper could not be made a gaoler for ever; and Essex was sent into his own house, after Anthony Bacon and all his other friends had been turned out of it, to be kept there in as close confinement as before. Then her mind again moved towards the Star Chamber. She summoned Bacon, who once more advised against it; once more he told her—not that the Earl's misdoings hardly deserved so terrible a form of prosecution—but that his power with the people was such as to make it dangerous. This time she agreed with him, and decided to set up a disciplinary tribunal of her own devising. There should be a fine show, and the miscreant should be lectured, very severely lectured, made to apologise, frightened a little, and then—let off. So she arranged it, and every one fell in with her plans. Never was the cool paternalism of the Tudors so curiously displayed. Essex was a naughty boy, who had misbehaved, been sent to his room, and fed on bread and water; and now he was to be brought downstairs, and, after a good wigging, told he was not to be flogged after all.

The ceremony took place (June 5th, 1600) at York House, and lasted for eleven hours without a break. Essex knelt at the foot of the table, round which the assembled lords of the Council sat in all their gravity. After some time the Archbishop of Canterbury moved that the Earl be allowed to stand; this was granted; later on he was allowed to lean, and at last to sit. The crown lawyers rose one after another to denounce his offences, which, with a few additions, were those specified in the Star Chamber declaration. Among the accusers was Bacon. He had written an ingenious letter, begging to be excused from taking a part in the proceedings, but adding that, if Her Majesty desired it, he could not refuse. Naturally enough, Her Majesty did desire it, and Bacon was instructed to draw the attention of the lords to the Earl's impropriety in accepting the dedication of Hayward's history of Henry IV. He knew full well the futility of the charge, but he did as he was bid. All was going well, and Essex was ready with a profound apology, when the dignity of the scene was marred by the excited ill-humour of Edward Coke, the Attorney-General. Essex found himself being attacked in such a way that he could not refrain from angry answers; Coke retorted; and the proceedings were degenerating into a wrangle, when Cecil intervened with some tactful observations. Then the judgment of the Court was given. Imprisonment in the Tower and an enormous fine were hung for a moment over the Earl's head; but on his reading aloud an abject avowal of his delinquencies, followed by a prayer for mercy, he was told that he might return to his house, and there await the Queen's pleasure.

He waited for a month before anything happened; at last his guards were removed, but he was still commanded to keep to his house. Not until the end of August was he given complete liberty. Elizabeth was relenting, but she was relenting as unpleasantly as possible. All through the summer she was in constant conference with Bacon, who had now taken up the *rôle* of intermediary between the Queen and the Earl. He had sent an apology to Essex for the part he had played at York House, and Essex had magnanimously accepted it. He now composed two elaborate letters, in Essex's name, addressed to the Queen and imploring her forgiveness. He did more. He invented a letter from his brother Anthony to Essex and the Earl's reply—brilliant compositions, in which the style of each was exquisitely imitated, and in which the Earl's devotion to his sovereign was beautifully displayed; and then he took these works and showed them to the Queen. Incidentally, there was much in them to the credit of Francis Bacon; but their effect was small. Perhaps Elizabeth was too familiar with the stratagems of plotters in the theatre to be altogether without suspicions when they were repeated in real life.

But Essex was not dependent upon Bacon's intervention; he wrote to the Queen himself, again and again. In varying tones he expressed his grief, he besought for an entire forgiveness, he begged to be allowed into the beloved presence once more. "Now having heard the voice of your Majesty's justice, I do humbly crave to hear your own proper and natural voice of grace, or else that your Majesty in mercy will send me into another world." "I receive no grace, your Majesty shows no mercy. But if your Majesty will vouchsafe to let me once prostrate myself at your feet and behold your fair and gracious eyes, though it be unknown to all the world but to him that your Majesty shall appoint to bring me to that paradise—yea, though afterwards your Majesty punish me, imprison me, or pronounce the sentence of death against me—your Majesty is most merciful, and I shall be most happy." So he wrote, but it was not only to Elizabeth that he addressed himself. Even while he was pouring out these regrets and protestations, his mind kept reverting to Ireland. One day he sent for Sir Charles Davers and asked him to make yet one more attempt upon the fidelity of Mountjoy. Davers knew well enough how it would be; but he was absolutely devoted to the Earl who, as he said afterwards, "had saved my life, and that after a very noble fashion; he had suffered for me, and made me by as many means bound unto him, as one man could be bound unto another; the life he had saved, and my estate and means whatsoever, he should ever dispose of"; and the adoring vassal immediately took horse to do as his lord desired.

A moment of crisis was approaching, which, Essex perceived, would reveal the real state of Elizabeth's mind. The monopoly of the sweet wines, which she had granted him for ten years, would come to an end at Michaelmas; would she renew it? It brought him a great income, and if she cut that off she would plunge him into poverty. Favour and hope—disgrace and ruin—those were the alternatives that seemed to hang upon her decision in this matter. She was well aware of it herself. She spoke of it to Bacon. "My Lord of Essex," she said, "has written me some very dutiful letters, and I have been moved by them; but" —she laughed grimly—"what I took for the abundance of the heart I find to be only a suit for the farm of sweet wines."

One letter, however, perhaps moved her more than the rest. "Haste, paper, to that happy presence, whence only unhappy I am banished; kiss that fair correcting hand which lays new plasters to my lighter hurts, but to my greatest wound applieth nothing. Say thou comest from pining, languishing, despairing ESSEX."

Did she find those words impossible to resist? It may have been so. From some phrases in another letter we may guess that there was indeed a meeting; but, if there was, it ended disastrously. In the midst of his impassioned speeches a fearful bitterness welled up within her; she commanded him from her presence; and with her own hands she thrust him out.[1]

She hesitated for a month, and then it was announced that the profits from the sweet wines would be henceforward reserved for the Crown. The effect upon Essex was appalling: he became like one possessed. Davers had already brought back word from Mountjoy that his decision was irreversible. "He desired my Lord to have patience, to recover again by ordinary means the Queen's ordinary favour; that, though he had it not in such measure as he had had heretofore, he should content himself." Patience! Content himself! The time for such words was past! He raved in fury, and then, suddenly recoiling, cursed himself in despair. "He shifteth," wrote Harington, who paid him at this time a brief and terrified visit, "from sorrow and repentance to rage and rebellion so suddenly as well proveth him devoid of good reason or right mind. . . . He uttered strange words, border-

[1] "This is but one of the many letters which, since I saw your Majesty, I wrote, but never sent unto you . . . I sometimes think of running [*i.e.* in the tiltyard] and then remember what it will be to come in armour triumphing into that presence, out of which both by your own voice I was commanded, and by your hands thrust out." Essex to the Queen. Undated.

ing on such strange designs that made me hasten forth and leave his presence. . . . His speeches of the Queen becometh no man who hath *mens sana in corpore sano*. He hath ill advisers and much evil hath sprung from this source. The Queen well knoweth how to humble the haughty spirit, the haughty spirit knoweth not how to yield, and the man's soul seemeth tossed to and fro, like the waves of a troubled sea."

His "speeches of the Queen" were indeed insane. On one occasion something was said in his presence of "Her Majesty's conditions." "Her conditions!" he exclaimed. "Her conditions are as crooked as her carcase!" The intolerable words reached Elizabeth and she never recovered from them.

She, too, perhaps was also mad. Did she not see that she was drifting to utter disaster? That by giving him freedom and projecting him into poverty, by disgracing him and yet leaving him uncrushed, she was treating him in the most dangerous manner that could be devised? Her life-long passion for half-measures, which had brought her all her glory, had now become a mania, and was about to prove her undoing. Involved in an extraordinary paralysis, she ignored her approaching fate.

But the Secretary ignored nothing. He saw what was happening, and what was bound to follow. He knew all about the gatherings at Lord Southampton's in Drury House. He noted the new faces come up from the country, the unusual crowds of swaggering gentlemen in the neighbourhood of the Strand, the sense of stir and preparation in the air; and he held himself ready for the critical moment, whenever it might come.

XIV

FOR Essex had now indeed abandoned himself to desperate courses. Seeing no more of Anthony Bacon, he listened only to the suggestions of his mother and Penelope Rich, to the loud anger of Sir Christopher Blount, and to the ruthless counsel of Henry Cuffe. Though Mountjoy had abandoned him, he still carried on a correspondence with the King of Scotland, and still hoped that from that direction deliverance might come. Early in the new year (1601) he wrote to James, asking him to send an envoy to London, who should concert with him upon a common course of action. And James, this time, agreed; he ordered the Earl of Mar to proceed to England, while he sent Essex a letter of encouragement. The letter arrived before the ambassador; and Essex preserved it in a small black leather purse, which he wore concealed about his neck.

The final explosion quickly followed. The Earl's partisans were seething with enthusiasm, fear, and animosity. Wild rumours were afloat among them, which they disseminated through the City. The Secretary, it was declared, was a friend to the Spaniards; he was actually intriguing for the Spanish

Infanta to succeed to the Crown of England. But more dangerous still was the odious Raleigh. Every one knew that that man's ambition had no scruples, that he respected no law, either human or divine; and he had sworn—so the story flew from mouth to mouth—to kill the Earl with his own hand, if there was no other way of getting rid of him. But perhaps the Earl's enemies had so perverted the mind of the Queen that such violent measures were unnecessary. During the first week of February the rumour rose that he was to be at once committed to the Tower. Essex himself perhaps believed it; he took counsel with his intimates; and it seemed to them that it would be rash to wait any longer for the arrival of Mar; that the time had come to strike, before the power of initiative was removed from them. But what was to be done? Some favoured the plan of an attack upon the Court, and a detailed scheme was drawn up, by which control was to be secured over the person of the Queen with a minimum of violence. Others believed that the best plan would be to raise the City in the Earl's favour; with the City behind them, they could make certain of overawing the Court. Essex could decide upon nothing; still wildly wavering, it is conceivable that, even now, he would have indefinitely postponed both projects and relapsed into his accustomed state of hectic impotence, if something had not happened to propel him into action.

That something bears all the marks of the gentle genius of Cecil. With unerring instinct the Secretary saw that the moment had now arrived at which it would be well to bring matters to an issue; and accordingly he did so. It was the faintest possible touch. On the morning of Saturday, February 7th, a messenger arrived from the Queen at Essex House, requiring the Earl to attend the Council. That was enough. To the conspirators it seemed obvious that this was an attempt to seize upon the Earl, and that, unless they acted immediately, all would be lost. Essex refused to move; he sent back a message that he was too ill to leave his bed; his friends crowded about him; and it was determined that the morrow should see the end of the Secretary's reign.

The Queen herself—who could be so base or so mad as to doubt it?—was to remain inviolate. Essex constantly asserted it; and yet there were some, apparently, among that rash multitude, who looked, even upon the divine Gloriana, with eyes that were profane. There was a singular episode on that Saturday afternoon. Sir Gilly Merrick, one of the most fiery of the Earl's adherents, went across the river with a group of his friends, to the players at Southwark. He was determined, he said, that the people should see that a Sovereign of England could be deposed, and he asked the players to act that afternoon the play of "Richard the Second." The players demurred: the play was an old one, and they would lose money by its performance. But Sir Gilly insisted; he offered them forty shillings if they would do as he wished; and on those terms the play was acted. Surely a strange circumstance! Sir Gilly must have been more conversant with history than literature; for how otherwise could he have imagined that the spectacle of the pathetic ruin of Shakespeare's minor poet of a hero could have nerved any man on earth to lift a hand, in actual fact, against so oddly different a ruler?

The Government, aware of everything, took its precautions, and on Sunday morning the guards were doubled at Whitehall. Sir Charles Davers went there early to reconnoitre, and returned with the news that it was no longer possible to surprise the Court; he recommended the Earl to escape secretly from London, to make his way into Wales, and there raise the standard of revolt. Sir Christopher Blount was for immediate action, and his words were strengthened by the ever-increasing crowd of armed men, who, since daybreak, had been pouring into the courtyard of Essex House. Three hundred were collected there by ten o'clock, and Essex was among them, when there was a knocking at the gate. The postern was opened, and four high dignitaries—the Lord Keeper, the Earl of Worcester, Sir William Knollys, and the Lord Chief Justice—made their appearance. Their servants were kept out, but they themselves were admitted. They had come, said Egerton, from the Queen, to enquire the cause of this assembly, and to say that if it arose through any grief against any persons whatsoever all complaints should be heard and justice be given. The noise and tumult were so great that conversation was impossible, and Essex asked the stately but agitated envoys to come up with him into his library. They did so, but hardly had they reached the room when the crowd burst in after them. There were cries of "Kill them! kill them!" and others of "Shop them up!" The Earl was surrounded by his shouting and gesticulating followers. He tried to speak, but they interrupted him. "Away, my Lord, they abuse you; they betray you; they undo you; you lose time!" He was powerless among them, and, while the Lords of the Council vainly adjured them to lay down their arms and depart in peace, he found himself swept towards the door. He bade Egerton and the others stay where they were; he would return ere long, he cried out, and go with them to the Queen. Then he was out of the room, and the door was shut and locked on the Councillors; they were "shopped up." Down the stairs and into the courtyard streamed the frenzied mob. And then the great gates were opened and they all rushed out into the street. But even now, at this last moment, there was hesitation. Where were they to go? "To the Court! To the Court!" cried some, and all waited upon Essex. But he, with a sudden determination, turned towards the City. To the City, then, it was to be. But there were no horses for such a multitude; they must all walk. The Strand lay before them, and down the Strand they hurried, brandishing their weapons. In front of all strode the tall black figure of Sir Christopher Blount. "Saw! Saw! Saw! Saw! Tray! Tray!" he shouted, seeking with wild gestures and incoherent exclamations to raise up London for the Earl.

The insurgents entered the City by Lud Gate; but the Government had been beforehand with them. Word had been sent to the preachers to tell the citizens to keep themselves within doors, armed, until further orders; and the citizens obeyed. Why should they do otherwise? The Earl was their hero; but they were loyal subjects of the Queen. They were quite unprepared for this sudden outbreak; they could not understand the causes of it; and then the news reached them that the Earl had been proclaimed a traitor; and the awful word and the ghastly penalties it carried with it struck terror into their souls. By noon Essex and his band were at St. Paul's, and there was no sign of any popular movement. He walked onward, crying aloud as he went that there was a plot to murder him, and that the Crown had been sold to the Spanish Infanta. But it was useless; there was no response; not a creature joined him. Those who were in the street stood still and silent, while perplexed and frightened faces peered out at him from doors and windows on either side. He had hoped to make a speech at Paul's Cross, but in such an atmosphere a set oration was clearly impossible; and besides, his self-confidence had now utterly gone. As he walked on down Cheapside, all men could see that he was in desperation; the sweat poured from his face, which was contorted in horror; he knew it at last—he was ruined—his whole life had crashed to pieces in this hideous fiasco.

In Gracechurch Street he entered the house of one of his friends, Sheriff Smith, upon whose support he reckoned. But the Sheriff, though sympathetic, was not disloyal, and he withdrew, on the pretext of consulting the Lord Mayor. After refreshing himself a little, Essex emerged, to find that many of his followers had slipped away, while the forces of the Government were gathering against him. He determined to return to his house; but at Lud Gate he found that the way was blocked. The Bishop of London and Sir John Leveson had collected together some soldiers and well-disposed citizens, and had stretched some chains across the narrow entry. The rebels charged, and were repelled. Sir Christopher was wounded; a page was killed; and some others were mortally injured. Essex turned down to the river. There he took boat, and rowed to Essex House, which he entered by the water-gate. The Councillors, he found, had been set free, and had returned to Whitehall. Having hurriedly destroyed a mass of incriminating papers, including the contents of the black leather purse about his neck, he proceeded to barricade the house. Very soon the Queen's troops, headed by the Lord Admiral, were upon him; artillery was brought up, and it was clear that resistance was useless. After a brief parley, Essex surrendered at discretion, and was immediately conveyed to the Tower.

XV

THE Government had never been in any danger, though there must have been some anxious moments at Whitehall. It was conceivable that the City might respond to the Earl's incitement and that a violent struggle would be the consequence; but Elizabeth, who was never lacking in personal courage, awaited the event with vigorous composure. When the news came that all was well, and she knew that she could depend upon the loyalty of the people, she found herself without a qualm. She gave orders that Essex and his adherents should be put upon their trial immediately.

Nearly a hundred persons were in custody, and the Council proceeded at once with the examination of the ringleaders. Very soon the whole course of the intrigues of the last eighteen months, including the correspondence with James and the connivance of Mountjoy, had transpired. The trial of the two Earls, Essex and Southampton, was fixed to take place before a special commission of Peers on February 18th. What line was the prosecution to take? It was speedily decided that no reference whatever should be made to Scotland, and that the facts incriminating

Mountjoy, whose services in Ireland could not be dispensed with, should be suppressed. There would be ample evidence of treason without entering upon such delicate and embarrassing particulars.

Bacon had been employed in the preliminary examination of some of the less important prisoners, and was now required to act as one of the counsel for the prosecution. He had no hesitations or doubts. Other minds might have been confused in such a circumstance; but he could discriminate with perfect clarity between the claims of the Earl and the claims of the Law. Private friendship and private benefits were one thing; the public duty of taking the part required of him by the State in bringing to justice a dangerous criminal was another. It was not for him to sit in judgment: he would merely act as a lawyer—merely put the case for the Crown, to the best of his ability, before the Peers. His own opinions, his own feelings, were irrelevant. It was true, no doubt, that by joining in the proceedings he would reap considerable advantages. From the financial point of view alone the affair would certainly be a godsend, for he was still pressingly in debt; and, besides that, there was the opportunity of still further ingratiating himself with the man who now, undoubtedly, was the most powerful personage in England—his cousin, Robert Cecil. But was that an argument for declining to serve? It was nonsense to suppose so. Because a lawyer was paid his fee did it follow that his motives were disreputable? There was, besides, one further complication. It was clear that it would be particularly useful for the Government to number Francis Bacon among its active supporters. The Earl had been his patron, and was his brother's intimate friend; and, if he was now ready to appear as one of the Earl's accusers, the effect upon the public, if not upon the judges, would be certainly great; it would be difficult to resist the conclusion that the case against Essex must be serious indeed since Francis Bacon was taking a share in it. If, on the other hand, he refused, he would undoubtedly incur the Queen's displeasure and run the risk of actual punishment; it might mean the end of his career. What followed? Surely only a simpleton would be puzzled into hesitation. The responsibility for the Government's acts lay with the Government; it was not for him to enquire into its purposes. And if, by doing his duty, he avoided disaster—so much the better! Others might be unable to distinguish between incidental benefits and criminal inducements: for him it was all as clear as day.

Never had his intellect functioned with a more satisfactory, a more beautiful, precision. The argument was perfect; there was, in fact, only one mistake about it, and that was that it had ever been made. A simpleton might have done better, for a simpleton might have perceived instinctively the essentials of the situation. It was an occasion for the broad grasp of common humanity, not for the razor-blade of a subtle intelligence. Bacon could not see this; he could not see that the long friendship, the incessant kindness, the high generosity, and the touching admiration of the Earl had made a participation in his ruin a deplorable and disgraceful thing. Sir Charles Davers was not a clever man; but his absolute devotion to his benefactor still smells sweet amid the withered corruptions of history. In Bacon's case such reckless heroism was not demanded; mere abstention would have been enough. If, braving the Queen's displeasure, he had withdrawn to Cambridge, cut down his extravagances, dismissed Jones, and devoted himself to those sciences which he so truly loved . . . but it was an impossibility. It was not in his nature or his destiny. The woolsack awaited him. Inspired with the ingenious grandeur of the serpent, he must deploy to the full the long luxury of his coils. One watches, fascinated, the glittering allurement; one desires in vain to turn away one's face.

A State Trial was little more than a dramatic formality. The verdict was determined beforehand by the administration, and every one concerned was well aware that this was so. Such significance as the proceedings had were of a political nature; they enabled those in power to give a public expression of their case against the prisoner—to lay before the world the motives by which they wished it to be supposed that they were actuated. In the present case there was no doubt whatever of the technical guilt of the accused. The Court of Peers had consulted the judges, who had pronounced that the conduct of Essex and his followers on Sunday the 8th, whatever their intentions may have been, in itself constituted treason, so that sentence might have been passed immediately a formal proof of that conduct had been made. But that a walk through the City should involve such fearful consequences would outrage public feeling; and it was the object of the prosecution to show that Essex had been guilty of a dangerous and deliberate conspiracy. The fact that the most serious feature in the case—the intrigue with the King of Scotland—was to be suppressed was a handicap for the Crown lawyers; but their position was an extremely strong one. The accused were allowed no counsel; their right of cross-examination was cut down to a minimum; and the evidence of the most important witnesses was given in the shape of depositions read aloud to the Court—depositions which had been extracted in the Tower, and which it was impossible to control or verify. On the whole, it seemed certain that with a little good management the prosecution would be able to blacken the conduct and character of the prisoners in a way which would carry conviction—in every sense of the word.

It so happened, however, that good management was precisely what was lacking on the part of the Crown leader, Edward Coke. On this far more serious occasion, the Attorney-General repeated the tactical errors which he had committed at York House. He abused his antagonists so roughly as to raise sympathy on their behalf; and he allowed himself to be led away into heated disputations which obscured the true issues of the case. During these wranglings, Essex was more than once able to carry the war into the enemy's camp. He declared fiercely that Raleigh had intended to murder him, and Raleigh was put into the witness-box to deny the irrelevant charge. A little later Essex brought up the story that the succession had been sold to the Spaniards by the Secretary. There followed a remarkable and unexpected scene. Cecil, who had been listening to the proceedings from behind a curtain, suddenly stepped forth, and, falling on his knees, begged to be allowed to clear himself of the slander. It was agreed that he should be heard, and, after a long altercation with Essex, Cecil elicited the fact that the informant upon whose report the charge was based was Sir William Knollys, the Earl's uncle. Knollys in his turn was sent for, and his evidence exculpated the Secretary. All that had happened, he said, was that Cecil had once mentioned to him a book in which the Infanta's title was preferred before any other. Essex's accusations had collapsed; but the prosecution, after many hours, had come no nearer to a proof of his criminal intentions. It was useless for Coke to shout and hector. "It was your purpose," he cried, shaking a menacing finger at Essex, "to take not only the Tower of London but the royal palace and the person of the prince—yea, and to take away her life!" Such exaggerations were only damaging to his own cause.

Bacon saw what was happening, and judged that it was time to intervene. The real question at issue—the precise nature of the Earl's motives—was indeed a complicated and obscure one. The motives of the most ordinary mortal are never easy to disentangle, and Essex was far from ordinary. His mind was made up of extremes, and his temper was devoid of balance. He rushed from opposite to opposite; he allowed the strangest contradictories to take root together, and grow up side by side, in his heart. He loved and hated—he was a devoted servant and an angry rebel—all at once. For an impartial eye, it is impossible to trace in his conduct a determined intention of any kind. He was swept hither and thither by the gusts of his passions and the accidents of circumstance. He entertained treasonable thoughts, and at last treasonable projects; but fitfully, with intervals of romantic fidelity and noble remorse. His behaviour in Ireland was typical of all the rest. After suggesting an invasion of England at the head of his troops, he veered completely round and led his army against Tyrone. It finally turned out that he had gone too far to draw back, and, pushed on by his own followers and the animosity of the Queen, he had plunged into a desperate action. But, till the last moment, he was uncertain, indefinite and distraught. There was no settled malignancy in his nature. It is possible that he believed in the treachery of Cecil; and, as it happened, there was some justification for the belief, for Cecil, with all his loyalty, was actually in receipt of a Spanish pension. Convinced of his own high purposes, the unrealistic creature may well have dreamed in his sanguine hours that, after all, he would manage to effect a bloodless revolution; that Cecil and Raleigh could be not too roughly thrust upon one side; and that then the way would be open once more for his true affection, his true admiration, his true ambition—that thenceforward the Queen would be his and he the Queen's, in glorious happiness, until death parted them.

Such were his inward workings, and Francis Bacon was the last man in the world to have understood them. They were utterly remote from the clear, bright ambit of that supremely positive intelligence. Wish as he might, the author of the "Essays or Counsels" could never have comprehended a psychology that was dominated by emotion instead of reason; but, on this occasion, he did not wish. Sympathy was far from him. What were the actual facts? By facts alone was it possible to judge of conduct, and the Court, led away by recrimination and irrelevancies, was beginning to lose sight of them. It was for him to brush aside, calmly but firmly, the excuses and the subterfuges of the prisoner, and to concentrate the attention of the judges—and of the public—on what was really the vital point in the whole business—the meaning of his deeds.

With perfect tact Bacon paid homage to the education of the Peers by illustrating his remarks with an incident from the Classics. All history, he said, made it plain "that there was never any traitor heard of, but he always coloured his practices with some plausible pretence." Essex had "made his colour the severing of some great men and counsellors from her Majesty's favour, and the fear he stood in of his pretended enemies lest they should murder him in his house. Therefore he saith he was compelled to fly into the City for succour and assistance." He was "not much unlike Pisistratus, of whom it was so anciently written how he gashed and wounded himself, and in that sort ran crying into Athens that his life was sought and like to have been taken away; thinking to have moved the people to have pitied him and taken his part by such counterfeited harm and danger: whereas his aim and drift was to take the government of the city into his hands, and alter the form thereof.

With like pretences of dangers and assaults, the Earl of Essex entered the City of London." In reality "he had no such enemies, no such dangers." The facts were plain, "and, my Lord"—he turned to the prisoner—"all whatsoever you have or can say in answer hereof are but shadows. And therefore methinks it were best for you to confess, not to justify."

Essex could never distinguish very clearly between a personality and an argument. "I call forth Mr. Bacon," he replied, "against Mr. Bacon"; and then he told the Court how, but a few months previously, his accuser had written letters in his name, to be shown to the Queen, in which his case had been stated "as orderly for me as I could do myself." "These digressions," said Bacon coldly, "are not fit, neither should be suffered;" the letters were harmless; "and," he added, "I have spent more time in vain in studying how to make the Earl a good servant to the Queen and state than I have done in anything else."

Then he sat down, and the case came once more under the guidance of Coke. The confessions of the other conspirators were read; but there was no order in the proceedings; point after point was taken up and dropped; and, at last, when the Attorney-General, after an harangue on the irreligion of the accused, offered to produce evidence upon the subject, the Peers declined to hear it. Once more confusion had descended, and once more Bacon rose to fix attention upon the central issue. "I have never yet seen in any case such favour shown to any prisoner," he said, "so many digressions, such delivery of evidence by fractions, and so silly a defence of such great and notorious treasons." He then read aloud the opinion of the judges on the point of law, and continued: "To take secret counsel, to execute it, to run together in numbers armed with weapons—what can be the excuse? Warned by the Lord Keeper, by a herald, and yet persist. Will any simple man take this to be less than treason?" Essex interrupted. "If I had purposed anything against others than my private enemies," he said, "I would not have stirred with so slender a company." Bacon paused a moment and then replied, addressing himself directly to the Earl. "It was not the company you carried with you, but the assistance which you hoped for in the City, which you trusted unto. The Duke of Guise thrust himself into the streets of Paris, on the day of the Barricadoes, in his doublet and hose, attended only with eight gentlemen, and found that help in the city, which (God be thanked) you failed of here. And what followed? The King was forced to put himself into a pilgrim's weeds, and in that disguise to steal away to escape their fury. Even such," he concluded, turning to the Peers, "was my Lord's confidence too; and his pretence the same—an all-hail and a kiss to the City.

But the end was treason, as hath been sufficiently proved."

The thrust was indeed a sharp one; but Bacon's words were no longer directed merely to the Court and the public. The parallel with Guise, whose rebellion had occurred within living memory, had in it an actuality far more deadly than the learned allusion to Pisistratus. There could be only one purpose in drawing it: it was precisely calculated to touch, in the most susceptible place, the mind of the Queen. To put Essex before her, with such verisimilitude, in the shape of the man who had raised up Paris against Henry III, was a master-stroke of detraction. The words, no doubt, would reach Elizabeth; but they were addressed, in reality, to some one else—to the invisible listener, who, after his dramatic appearance, had returned to his place behind the hangings. The Secretary's kindred intellect appreciated to the full the subtle implications of the speech; his cousin was doing admirably. The Earl was silent. Francis Bacon's task was over. The double tongue had struck, and struck again.

Both prisoners were inevitably found guilty, and the revolting sentence was passed in the usual form. During the ordeal of the trial Essex had been bold, dignified, and self-possessed; but now, back again in the Tower, he was seized by a violent revulsion; anguish and horror overpowered his mind. A puritan clergyman, who had been sent to minister to him, took the opportunity to agitate his conscience and fill his imagination with the fear of hell. He completely collapsed. Self-reliance—self-respect—were swept away in a flood of bitter lamentations. He wished, he said, to make a confession to the Lords of the Council. They came, and he declared to them that he was a miserable sinner, grovelling heart-broken before the judgment-seat of God. He cried out upon his inexcusable guilt; and he did more: he denounced the black thoughts, the fatal counsels, the evil doings of his associates. They, too, were traitors and villains, no less than himself. He raved against them all—his stepfather—Sir Charles Davers—Henry Cuffe—each was worse than the other; they had lured him on to these abominable practices, and now they were all to sink together under a common doom. His sister, too! Let her not be forgotten—she had been among the wickedest! Was she not guilty of more sins than one? "She must be looked to," he cried, "for she hath a proud spirit!"—adding dark words of Mountjoy, and false friendship, and broken vows of marriage. Then, while the grave Councillors listened in embarrassed silence, he returned once more to his own enormities. "I know my sins," he said, "unto her Majesty and to my God. I must confess to you that I am the greatest, the most vilest and most unthankful traitor that has ever been in the land."

While these painful scenes of weakness and humiliation were passing in the Tower, Elizabeth had withdrawn into deepest privacy at Whitehall. Every mind was turned towards her—in speculation, in hope, in terror; the fatal future lay now, spinning and quivering, within her formidable grasp.

It is not difficult to guess the steps by which she reached her final conclusion. The actual danger which she had run must have seemed to her—in spite of Bacon's reminder—the least important element in the case. The rising had been an act of folly, doomed from the first to ignominious failure—an act so weak and ineffective that, taken by itself, it could hardly be said to deserve the extreme penalty of the Law. If, for other reasons, she was inclined towards mercy, there would be ample justification for taking a lenient view of what had happened, and for commuting the punishment of death for one, perhaps, of imprisonment and sequestration. It is true that the intrigue with James of Scotland wore a more serious complexion; but this had proved abortive; it was unknown to all but a few in high places; and it might well be buried in oblivion. Were there, then, other reasons for mercy? Most assuredly there were. But these were not judicial reasons; neither were they political; they were purely personal; and, of course, in that very fact lay their strength.

To abolish, in a moment, the immediate miserable past—to be reconciled once more; to regain, with a new rapture, the old happiness—what was there to prevent it? Nothing, surely; she had the power for such an act; she could assert her will—extend her royal pardon; after a short eclipse, he would be with her again; not a voice would be raised against her; Cecil himself, she knew, would accept the situation without a murmur; and so—would not all be well? It was indeed a heavenly vision, and she allowed herself to float deliciously down the stream of her desires. But not for long. She could not dwell indefinitely among imaginations; her sense of fact crept forward—insidious—paramount; with relentless fingers it picked to pieces the rosy palaces of unreality. She was standing once again on the bleak rock. She saw plainly that she could never trust him, that the future would always repeat the past, that, whatever her feelings might be, his would remain divided, dangerous, profoundly intractable, and that, if this catastrophe were exorcised, another, even worse, would follow in its place.

And yet, after all, might she not take the risk? She had been a gambler all her life; there was little left of it now; why not live out that little in the old style, with the old hazard—the close-hauled boat tacking fiercely against the wind? Let him intrigue with James of Scotland, she could manage that! Let him do his worst—she would be equal to it; she would wrestle with him, master him, hold him at her mercy, and pardon him—magnificently, ecstatically, pardon him—again and again! If she failed, well, that would be a new experience, and—how often had she said it! —*per molto variare la natura è bella.* Yes, truly, she and nature were akin—variable, beautiful . . . a hideous memory struck her; terrible outrageous words re-echoed in her mind. "Crooked"—"carcase"—so that was what he thought of her! While he was pouring out his sugared adorations, he loathed her, despised her, recoiled from her. Was it possible? Was the whole history of their relations, then, one long infamous deception? Was it all bitterness and blindness? Had he perhaps truly loved her once?—Once! But the past was over, and time was inexorable. Every moment widened the desperate abyss between them. Such dreams were utter folly. She preferred not to look in her looking-glass—why should she? There was no need; she was very well aware without that of what had happened to her. She was a miserable old woman of sixty-seven. She recognised the truth—the whole truth—at last.

Her tremendous vanity—the citadel of her repressed romanticism—was shattered, and rage and hatred planted their flag upon its ruins. The animosity which for so long had been fluctuating within her now flared up in triumph and rushed out upon the author of her agony and her disgrace. He had betrayed her in every possible way—mentally, emotionally, materially—as a Queen and as a woman—before the world and in the sweetest privacies of the heart. And he had actually imagined that he could elude the doom that waited on such iniquity—had dreamed of standing up against her—had mistaken the hesitations of her strength for the weaknesses of a subservient character. He would have a sad awakening! He would find that she was indeed the daughter of a father who had known how to rule a kingdom and how to punish the perfidy of those he had loved the most. Yes, indeed, she felt her father's spirit within her; and an extraordinary passion moved the obscure profundities of her being, as she condemned her lover to her mother's death. In all that had happened there was a dark inevitability, a ghastly satisfaction; her father's destiny, by some intimate dispensation, was repeated in hers; it was supremely fitting that Robert Devereux should follow Anne Boleyn to the block. Her father! . . . but in a still remoter depth there were still stranger stirrings. There was a difference as well as a likeness; after all, she was no man, but a woman; and was this, perhaps, not a repetition but a revenge? After all the long years of her life-time, and in this appalling consummation, was it her murdered mother who had finally emerged? The wheel had come full circle. Manhood—the fascinating, detestable entity, which had first come upon her concealed in yellow magnificence in her father's lap—manhood was overthrown at last, and in the person of that traitor it should be rooted out. Literally, perhaps . . . she knew well enough the punishment for high treason. But no! She smiled sardonically. She would not deprive him of the privilege of his rank. It would be enough if he suffered as so many others—the Lord Admiral Seymour among the rest—had suffered before him; it would be enough if she cut off his head.

And so it happened that this was the one occasion in her life on which Elizabeth hardly hesitated. The trial had taken place on February the 19th, and the execution was fixed for the 25th. A little wavering there had indeed to be—she would not have been Elizabeth without it; but it was hardly perceptible. On the 23rd she sent a message that the execution should be postponed; on the 24th she sent another that it should be proceeded with. She interfered with the course of the law no further.

Afterwards a romantic story was told, which made the final catastrophe the consequence of a dramatic mishap. The tale is well known: how, in happier days, the Queen gave the Earl a ring, with the promise that, whenever he sent it back to her, it would always bring forgiveness; how Essex, leaning from a window of the Tower, entrusted the ring to a boy, bidding him take it to Lady Scrope, and beg her to present it to her Majesty; how the boy, in mistake, gave the ring to Lady Scrope's sister, Lady Nottingham, the wife of the Earl's enemy; how Lady Nottingham kept it, and said nothing, until, on her deathbed two years later, she confessed all to the Queen, who, with the exclamation "God may forgive you, Madam, but I never can!" brought down the curtain on the tragedy. Such a narrative is appropriate enough to the place where it was first fully elaborated—a sentimental novelette;[1] but it does not belong to history. The improbability of its details is too glaring, and the testimony against it is overpowering. It is implicitly denied by Camden, the weightiest of contemporary historians; it is explicitly contradicted by Clarendon, who, writing in the succeeding generation, was in a position to know the facts; and it has been rejected by later writers, including the learned and judicious Ranke. And assuredly the grim facts stand better by themselves, without the aid of such adventitious ornaments. Essex made no appeal. Of what use would be a cry for mercy? Elizabeth would listen to nothing, if she was deaf to her own heart. The end came in silence: and at last he understood. Like her other victims, he realised too late that he had utterly misjudged her nature, that there had never been the slightest possibility of dominating her, that the enormous apparatus of her hesitations and collapses was merely an incredibly elaborate façade, and that all within was iron.

[1] *The Secret History of the most renowned Queen Elizabeth and the Earl of Essex, by a Person of Quality*, 1695. A reference to the legend in its rudimentary form occurs in *The Devil's Law Case* (*circa* 1620). Cf. *The Works of John Webster*, ed. Lucas, ii. 343.

One request he made—that he should not be executed in public; and it was willingly granted, for there still seemed a chance of a popular movement on his behalf. He should be beheaded, like all the great state criminals before him, in the courtyard of the Tower.

And there, on the morning of February 25th, 1601, were gathered together all those who were qualified to witness the closing ceremony. Among them was Walter Raleigh. As Captain of the Guard, it was his duty to be present; but he had thought, too, that perhaps the condemned man would have some words to say to him, and he took up his station very near the block. There were murmurs around him. Was this as it should be? Now that the great Earl was brought so low, were his enemies to come pressing about him in scornful jubilation? A shameful sight! Raleigh heard, and in sombre silence immediately withdrew He went into the White Tower, ascended to the Armoury, and thence, from a window, the ominous prophet of imperialism surveyed the scene.

It was not a short one. The age demanded that there should be a dignified formality on such occasions, and that the dreadful physical deed should be approached through a long series of ornate and pious commonplaces. Essex appeared in a black cloak and hat with three clergymen beside him. Stepping upon the scaffold, he took off his hat, and bowed to the assembled lords. He spoke long and earnestly—a studied oration, half speech, half prayer. He confessed his sins, both general and particular. He was young, he said—he was in his thirty-fourth year—and he "had bestowed his youth in wantonness, lust, and uncleanness." He had been "puffed up with pride, vanity, and love of this world's pleasure"; his sins were "more in number than the hairs on his head." "For all which," he went on, "I humbly beseech my saviour Christ to be a mediator to the eternal Majesty for my pardon; especially for this my last sin, this great, this bloody, this crying, this infectious sin, whereby so many for love of me have been drawn to offend God, to offend their sovereign, to offend the world. I beseech God to forgive it us, and to forgive it me—most wretched of all." He prayed for the welfare of the Queen, "whose death I protest I never meant, nor violence to her person." He was never, he declared, either an atheist or a papist, but hoped for salvation from God only by the mercy and merits of "my saviour Jesus Christ. This faith I was brought up in; and herein I am ready to die; beseeching you all to join your souls with me in prayer." He paused, and was about to take off his cloak, when one of the clergymen reminded him that he should pray God to forgive his enemies. He did so, and then, removing his cloak and ruff, knelt down by the block in his black doublet. Another of the clergymen encouraged him against the fear of death, whereupon, with ingenuous gravity, he confessed that more than once, in battle, he had "felt the weakness of the flesh, and therefore in this great conflict desired God to assist and strengthen him." After that, gazing upwards, he prayed, more passionately, to the Almighty. He prayed for all the Estates of the Realm, and he repeated the Lord's prayer. The executioner, kneeling before him, asked for his forgiveness, which he granted. The clergymen requested him to rehearse the creed, and he went through it, repeating it after them clause by clause. He rose and took off his doublet; a scarlet waistcoat, with long scarlet sleeves, was underneath. So—tall, splendid, bare-headed, with his fair hair about his shoulders—he stood before the world for the last time. Then, turning, he bowed low before the block; and, saying that he would be ready when he stretched out his arms, he lay down flat upon the scaffold. "Lord, be merciful to thy prostrate servant!" he cried out, and put his head sideways upon the low block. "Lord, into thy hands I recommend my spirit." There was a pause; and all at once the red arms were seen to be extended. The headsman whirled up the axe, and crashed it downwards; there was no movement; but twice more the violent action was repeated before the head was severed and the blood poured forth. The man stooped, and, taking the head by the hair, held it up before the onlookers, shouting as he did so, "God save the Queen!"

XVI

SOUTHAMPTON was spared. His youth and romantic devotion to Essex were accepted as a palliation of his delinquency, and the death sentence was commuted for imprisonment in the Tower. Sir Christopher Blount and Sir Charles Davers were beheaded; Sir Gilly Merrick and Henry Cuffe were hanged. Some heavy fines were levied from some of the other conspirators, but there were no more executions; the Government was less vindictive than might have been expected. Penelope Rich, who had been taken prisoner in Essex House at the same time as her brother, was set free. In the hour of his triumph, Cecil's one wish was to show no animosity; he gave rein to his instinctive mildness, and was as polite as possible to his fallen enemies. An opportunity occurred of showing a favour to Lady Essex, and he immediately seized it. One Daniell, a servant of the Earl's, had got hold of some of his private letters, had forged copies of them, and had blackmailed the Countess with threats of publication. She appealed to Cecil, who acted with great promptitude. The ruffian was seized and brought before the Star Chamber; and, in an elaborate sentence, filled with flowery praises of the Countess, he was condemned to pay her two thousand pounds, to be fined another thousand, to be imprisoned for life, and—"to thend the said offences of the foresaid Daniell should not only be notefyed to the publique viewe, but to cause others to refrayne from committing of the like hereafter, it is likewise ordered and decreed that for the same his offences he the said Daniell shalbe sett upon the pillory, with his eares thereunto nayled, with a paper on his head inscribed with these words—For forgery, corrupte cosenages, and other lewde practises." Lady Essex was duly grateful; a letter of thanks to Cecil gives us a momentary glimpse of the most mysterious of the personages in this tragic history. A shrouded figure, moving dubiously on that brilliantly lighted stage, Frances Walsingham remains utterly unknown to us. We can only guess, according to our fancy, at some rare beauty, some sovereign charm—and at one thing more: a super-abundant vitality. For, two years later, the widow of Sidney and Essex was married for the third time—to the Earl of Clanricarde. And so she vanishes.

The rising had been followed by no repercussions among the people, but the Government remained slightly uneasy. It was anxious to convince the public that Essex had not been made a martyr to political intrigue, but was a dangerous criminal who had received a righteous punishment. The preacher in St. Paul's was instructed to deliver a sermon to that effect, but this was not enough; and it was determined to print and publish a narrative of the circumstances, with extracts from the official evidence attached. Obviously Bacon was the man to carry out the work; he was instructed to do so; his labours were submitted to the correction of the Queen and Council; and the "Declaration of the Practices and Treasons of Robert Late Earl of Essex and his Complices . . . together with the very Confessions, and other parts of the Evidences themselves, word for word taken out of the Originals"[2] was the result. The tract was written with brevity and clarity, and, as was to be expected, it expressed in a more detailed form the view of the case which Bacon had outlined in his speeches at the trial. It showed that the rising had been the result of a long-thought-out and deliberately planned conspiracy. This result was achieved with the greatest skill and neatness; certain passages in the confessions were silently suppressed; but the manipulations of the evidence were reduced to a minimum; and there was only one actually false statement of fact. The date of the Earl's proposal to invade England with the Irish army was altered; it was asserted to have taken place after the expedition against Tyrone, and not before it; and thus one of the clearest indications of the indeterminate and fluctuating nature of Essex and his plans was not only concealed but converted into a confirmation of Bacon's thesis. By means of a clever series of small omissions from the evidence, the balance of the facts just previous to the rising was entirely changed; the Earl's hesitations—which in truth continued up to the very last moment—were obliterated, and it was made to appear that the march into the City had been steadily fixed upon for weeks. So small and subtle were the means by which Bacon's end was reached that one cannot but wonder whether, after all, he was conscious of their existence. Yet such a beautiful economy—could it have arisen unbeknownst? Who can tell? The serpent glides off with his secret.

As a reward for his services Francis Bacon received £1200 from the Queen. And very soon his financial position was improved still further. Three months after the final catastrophe, Anthony Bacon found the rest which this world had never given him. The terrible concatenation of events—the loss of his master, the loss of his brother, the ruin of his hopes, the triumph of folly, passion, and wickedness—had broken the last prop of his shattered health—his fierce indomitable spirit. He died, and Francis inherited his small fortune. The future was brightening. Property—prosperity—a multitude of satisfactions, sensual and intellectual—a crowded life of brilliance, learning, and power—were these things coming then at last? Perhaps; but when they came they would be shared in no family rejoicings. Only a strange cackle disturbed the silence of Gorhambury. For old Lady Bacon's wits had finally turned. Gibbering of the Lord and the Earl, of her sons and her nephew, of hell-fire and wantonness, she passed the futile days in a confusion of prayers and rages. Frantic, she tottered on into extreme senility. Oblivion covers her.

Mastery had come into Robert Cecil's hands; but it was mastery tempered by anxiety and vigilance. No sooner was his great rival gone than a fresh crisis, of supreme importance in his life, was upon him. The Earl of Mar arrived in London. The situation had completely changed since his departure from Scotland, and it now seemed as if James's emissary could have little to do at the English Court. While he was waiting indecisively, he received a message from Cecil, asking for a private interview. The Secretary had seen where the key to the future lay. He was able to convince Mar that he was sincerely devoted to the cause of the King of Scotland. If only, he said, James would abandon his policy of protests and clandestine manoeuvring, if he would put his trust in him, if he would leave to him the management of the necessary details, he would find, when the hour struck, that all would be well, that the transition would be accomplished and the crown of England his, without the slightest difficulty or danger. Mar, deeply impressed, returned to Edinburgh, and succeeded in making James understand the crucial importance of these advances. A secret correspondence began between the King and the Secretary. The letters, sent round, by way of precaution, through an intermediary in Dublin, brought James ever more closely under the wise and gentle sway of Cecil. Gradually, persistently, infinitely quietly, the obstacles in the path of the future were smoothed away; and the royal gratitude grew into affection, into devotion, as the inevitable moment drew near.

To Cecil, while he watched and waited, one possibility was more disturbing than all the rest. The rise of Raleigh had accompanied the fall of Essex; the Queen had made him Governor of Jersey; she was beginning to employ him in diplomacy; where was this to end? Was it conceivable that the upshot of the whole drama was merely to be a change of dangerous favourites—but a change for the worse, by which the dashing incompetence of Essex would be replaced by Raleigh's sinister force? And, even if it was too late now for that bold man to snatch very much more from Elizabeth, what fatal influence might he not come to wield over the romantic and easily impressible James? This must be looked to; and looked to it was. The King's mind was satisfactorily infected with the required sentiments; Cecil himself said very little—only a sharp word, once; but Lord Henry Howard, who, as Cecil's closest ally, had been allowed to join in the secret correspondence, poured out, in letter after letter, envenomed warnings and bitter accusations; and soon James felt for Raleigh only loathing and dread. Raleigh himself was utterly unsuspecting; there seemed to be a warm friendship between him and the Secretary. Once again he was the victim of bad luck. His earlier hopes had been shattered by Essex; and now that Essex was destroyed he was faced by a yet more dangerous antagonist. In reality, the Earl's ruin, which he had so virulently demanded, was to be the prologue of his own. As he had looked out from the armoury on his enemy's execution, his eyes had filled with tears. So strangely had he been melted by the grandeur of the tragedy! But did some remote premonition also move him? Some obscure prevision of the end that would be his too, at last?

The great reign continued for two years longer; but the pulses of action had grown feeble; and over public affairs there hovered a cloud of weariness and suspense. Only in one quarter was history still being made—in Ireland. Elizabeth's choice of Mountjoy had been completely justified. With relentless skill and energy he had worn down the forces of Tyrone. In was in vain that all Catholic Europe prayed for the rebel, in vain that the Pope sent him a phoenix's feather, in vain that three thousand Spaniards landed at Kinsale. Mountjoy was victorious in a pitched battle; the Spaniards were forced to capitulate;

Tyrone was pressed back, pursued, harried, driven from pillar to post. Once more he negotiated and yielded; but this time the dream of a Catholic dominion in Ireland was finally shattered, and Elizabeth's crowning triumph was achieved. Yet Tyrone's strange history was not ended; some unexpected sands were still waiting for him in Time's glass. A great lord once again on his estates in Ulster, rich and proud with his adoring vassals about him, he suddenly plunged into a fresh quarrel with the English Government. All at once he took fright—he fled. For long he wandered with his family and retinue through France, Flanders, and Germany, a desperate exile, an extraordinary flitting focus of ambiguous intrigue. At length the Pope received him, housed him, pensioned him; his adventures silently ceased. And he, too, passes from us—submerged by the long vague years of peace, indolence, and insignificance—sinking away into forgetfulness through the monotony of Roman afternoons.

Elizabeth had resisted the first onslaughts of rage and grief with the utmost bravery, but an inevitable reaction followed, and, as the full consciousness of what had happened pressed in upon her, her nervous system began to give way. Her temper grew more abrupt and capricious than ever; for days at a time she sat silent in moody melancholy. She could hardly bring herself to eat; "little but manchet and succory potage"—so Sir John Harington tells us—passed her lips. She kept a sword continually by her, and when a nerve-storm came upon her she would snatch it up, stamp savagely to and fro, and thrust it in fury into the tapestry. Sir John, when he begged for an audience, received a sharp reply. "Go tell that witty fellow, my godson, to get home; it is no season now to fool it here." It was too true, and he obeyed her, sad at heart. Sometimes she would shut herself up in a darkened room, in paroxysms of weeping. Then she would emerge, scowling, discover some imagined neglect, and rate her waiting-women until they, too, were reduced to tears.

She still worked on at the daily business of Government, though at times there were indications that the habits of a lifetime were disintegrating, and she was careless, or forgetful, as she had never been before. To those who watched her, it almost seemed as if the inner spring were broken, and that the mechanism continued to act by the mere force of momentum. At the same time her physical strength showed signs of alarming decay. There was a painful scene when, in October, she opened Parliament. As she stood in her heavy robes before the Lords and Commons, she was suddenly seen to totter; several gentlemen hurried forward and supported her; without them, she would have fallen to the ground.

But in truth the old spirit was not yet extinct, and

she was still capable of producing a magnificent sensation. The veteran conjurer's hand might tremble, but it had not lost the art of bringing an incredible rabbit out of a hat. When the session of Parliament began, it was found that there was great and general discontent on the subject of monopolies. These grants to private persons of the sole right to sell various articles had been growing in number, and were felt to be oppressive. As the long list of them was being read aloud in the House of Commons, a member interjected. "Is not bread there?" "If order be not taken," another replied, "it will be, before next Parliament." The monopolies—Essex's lease of the sweet wines had been one of them—were Elizabeth's frugal method of rewarding her favourites or officials; and to protest against them amounted to an indirect attack on the royal prerogative. Elizabeth had not been accustomed to put up with interferences of this kind from the Commons; how often, for less cause than this, had she railed at them in high displeasure, and dismissed them cowering from her presence! And so no one was surprised when she sent for the Speaker, and the poor man prepared himself for a tremendous wigging. Great was his amazement. She greeted him with the highest affability; told him that she had lately become aware that "divers patents, which she had granted, were grievous to her subjects," assured him that she had been thinking of the matter "even in the midst of her most great and weighty occasions,"[2] and promised immediate reform. The Speaker departed in raptures. With her supreme instinct for facts, she had perceived that the debate in the House represented a feeling in the country with which it would be unwise to come into conflict; she saw that policy dictated a withdrawal; and she determined to make the very best use of an unfortunate circumstance. The Commons were overwhelmed when they learnt what had happened; discontent was turned to adoration; there was a flood of sentiment, and the accumulated popularity of half a century suddenly leapt up to its highest point. They sent a deputation to express their gratitude, and she received them in state. "In all duty and thankfulness," said the Speaker, as the whole company knelt before her, "prostrate at your feet, we present our most loyal and thankful hearts, and the last spirit in our nostrils, to be poured out, to be breathed up, for your safety."[3] There was a pause; and then the high voice rang out:—"Mr. Speaker, we perceive your coming is to present thanks unto us; know I accept them with no less joy than your loves can have desired to offer such a present, and do more esteem it than any treasure or riches, for those we know how to prize, but loyalty, love, and thanks I account them invaluable; and, though God hath raised me high, yet this I account the glory of my crown, that I have reigned with your loves."[2] She stopped, and told them to stand up, as she had more to say to them. "When I heard it," she went on, "I could give no rest unto my thoughts until I had reformed it, and those varlets, lewd persons, abusers of my bounty, shall know I will not suffer it. And, Mr. Speaker, tell the House from me that I take it exceeding grateful that the knowledge of these things have come unto me from them. Of myself, I must say this, I never was any greedy scraping grasper, nor a strict fast-holding prince, nor yet a waster; my heart was never set upon any worldly goods, but only for my subjects' good." Pausing again for a moment, she continued in a deeper tone. "To be a king and wear a crown is a thing more glorious to them that see it than it is pleasant to them that bear it. The cares and troubles of a crown I cannot more fitly resemble than to the drugs of a learned physician, perfumed with some aromatical savour, or to bitter pills gilded over, by which they are made more acceptable or less offensive, which indeed are bitter and unpleasant to take. And for my own part, were it not for conscience' sake to discharge the duty that God hath laid upon me, and to maintain His glory, and keep you in safety, in mine own disposition I should be willing to resign the place I hold to any other, and glad to be freed of the glory with the labours; for it is not my desire to live nor to reign longer than my life and reign shall be for your good. And, though you have had and may have many mightier and wiser princes sitting in this seat, yet you never had nor shall have any love you better." She straightened herself with a final effort; her eyes glared; there was a sound of trumpets; and, turning from them in her sweeping draperies—erect and terrible—she walked out.

XVII

THE end approached very gradually—with the delay which, so it seemed, had become *de rigueur* in that ambiguous Court. The ordinary routine continued, and in her seventieth year the Queen transacted business, went on progress, and danced while ambassadors peeped through the hangings, as of old. Vitality ebbed slowly; but at times there was a sudden turn; health and spirits flowed in upon the capricious organism; wit sparkled; the loud familiar laughter re-echoed through Whitehall. Then the sombre hours returned again—the distaste for all that life offered—the savage outbursts—the lamentations. So it had come to this! It was all too clear—her inordinate triumph had only brought her to solitude and ruin. She sat alone, amid emptiness and ashes, bereft of the one thing in the whole world that was worth having. And she herself, with her own hand, had cast it from her, had destroyed it . . . but it was not true; she had been helpless—a puppet in the grasp of some malignant power, some hideous influence inherent in the very structure of reality. In such moods, with royal indifference, she unburdened her soul to all who approached her—to her ladies, to an ambassador, or to some old scholar who had come to show her his books. With deep sighs and mourning gestures she constantly repeated the name of Essex. Then she dismissed them—the futile listeners—with a wave of her hand. It was better that the inward truth should be expressed by the outward seeming; it was better to be alone.

In the winter of 1602, Harington came again to Court, and this time he obtained an audience of his godmother. "I found her," he told his wife, "in most pitiable state." Negotiations with Tyrone were then in progress, and she, forgetful of a former conversation, asked Sir John if he had ever seen the rebel. "I replied with reverence that I had seen him with the Lord Deputy; she looked up with much choler and grief in her countenance, and said, 'Oh, now it mindeth me that you was one who saw this man elsewhere,' and hereat she dropped a tear, and smote her bosom." He thought to amuse her with some literary trifles, and read her one or two of his rhyming epigrams. She smiled faintly. "When thou dost feel creeping time at thy gate," she said, "these fooleries will please thee less; I am past my relish for such matters."

With the new year her spirits revived, and she attended some state dinners. Then she moved to Richmond, for change of air; and at Richmond, in March, 1603, her strength finally left her. There were no very definite symptoms, besides the growing physical weakness and the profound depression of mind. She would allow no doctors to come near her; she ate and drank very little, lying for hours in a low chair. At last it was seen that some strange crisis was approaching. She struggled to rise, and failing, summoned her attendants to pull her to her feet. She stood. Refusing further help, she remained immovable, while those around her watched in awe-stricken silence. Too weak to walk, she still had the strength to stand; if she returned to her chair, she knew that she would never rise from it; she would continue to stand, then; had it not always been her favourite posture? She was fighting Death, and fighting with terrific tenacity. The appalling combat lasted for fifteen hours. Then she yielded—though she still declared that she would not go to bed. She sank on to cushions, spread out to receive her; and there she lay for four days and nights, speechless, with her finger in her mouth. Meanwhile an atmosphere of hysterical nightmare had descended on the Court. The air was thick with doom and terror. One of the ladies, looking under a chair, saw, nailed to the bottom of it, a queen of hearts. What did the awful portent mean? Another, leaving the Queen's room for a little rest, went down a gallery, and caught a glimpse of a shadowy form, sweeping away from her in the familiar panoply of Majesty. Distracted by fear, she retraced her steps, and, hurrying back into the royal chamber, looked—and beheld the Queen lying silent on the pillows, with her finger in her mouth, as she had left her.

The great personages about her implored her to obey the physicians and let herself be moved—in vain. At last Cecil said boldly, "Your Majesty, to content the people, you must go to bed." "Little man, little man," came the answer, "the word *must* is not used to princes." She indicated that she wished for music, and the instruments were brought into the room; with delicate melancholy they discoursed to her, and for a little she found relief. The consolations of religion remained; but they were dim formalities to that irretrievably terrestrial nature; a tune on the virginals had always been more to her mind than a prayer. Eventually she was carried to her bed. Cecil and the other Councillors gathered round her; had she any instructions, the Secretary asked, in the matter of her successor? There was no answer. "The King of Scotland?" he hinted; and she made a sign—so it seemed to him—which showed agreement. The Archbishop of Canterbury came—the aged Whitgift, whom she had called in merrier days her "little black husband"—and knelt beside her. He prayed fervently and long; and now, unexpectedly, she seemed to take a pleasure in his ministrations; on and on he prayed, until his old knees were in an agony, and he made a move as if to rise. But she would not allow it, and for another intolerable period he raised his petitions to heaven. It was late at night before he was released, when he saw that she had fallen asleep. She continued asleep, until —in the cold dark hours of the early morning of March 24th—there was a change; and the anxious courtiers, as they bent over the bed, perceived, yet once again, that the inexplicable spirit had eluded them. But it was for the last time: a haggard husk was all that was left of Queen Elizabeth.

But meanwhile, in an inner chamber, at his table, alone, the Secretary sat writing. All eventualities had been foreseen, everything was arranged, only the last soft touches remained to be given. The momentous transition would come now with exquisite facility. As the hand moved, the mind moved too, ranging sadly over the vicissitudes of mortal beings, reflecting upon the revolutions of kingdoms, and dreaming, with quiet clarity, of what the hours, even then, were bringing—the union of two nations—the triumph of the new rulers—success, power, and riches—a name in after-ages—a noble lineage—a great House

BIBLIOGRAPHY

Abbot, Edwin A. *Bacon and Essex*. 1877.

Abbot, Edwin A. *Francis Bacon*. 1885.

Aubrey, John. *Brief Lives*. Edited by Andrew Clark. 1898.

Bagenal, Philip H. *Vicissitudes of an Anglo-Irish Family*. 1925.

Birch, Thomas. *Memoirs of the Reign of Queen Elizabeth*. 1754.

Birch MSS, British Museum.

Brewer, J. S. *English Studies*. 1881.

Camden, William. *Rerum Anglicarum et Hibernicarum Annales, regnante Elizabetha*. 1625.

Cary, Robert. *Memoirs*. 1759.

Cecil, Sir Robert. *Letters to Sir George Carew*. Camden Society. 1864.

Chamberlain, John. *Letters*. Camden Society. 1861.

Chamberlin, Frederick. *The Private Character of Queen Elizabeth*. 1921.

Creighton, Mandell. *Queen Elizabeth*. 1906.

Devereux, Walter Bourchier. *Lives and Letters of the Devereux, Earls of Essex*. 1853.

Dictionary of National Biography.

Egerton Papers. Camden Society. 1840.

Elizabeth, Queen. *Correspondence with James VI of Scotland*. Camden Society. 1849.

English Historical Review. Vol. IX. 1894.

Froude, James Anthony. *History of England*. 1870.

Froude, James Anthony. *The Spanish Story of the Armada*. 1892.

Goodman, Godfrey. *The Court of King James the First*. 1839.

Gosse, Edmund. *The Life and Letters of John Donne*, 1899.

Harington, Sir John. *Nugae Antiquae*. 1779.

Index

Index

Index

Index

New Worlds
of Mind & Spirit
Eternal Growth